# Civic Ecology

**Urban and Industrial Environments**

Series editor: Robert Gottlieb, Henry R. Luce Professor of Urban and Environmental Policy, Occidental College

For a complete list of books published in this series, please see the back of the book.

# Civic Ecology
## Adaptation and Transformation from the Ground Up

Marianne E. Krasny and Keith G. Tidball

The MIT Press
Cambridge, Massachusetts
London, England

MIT Press books may be purchased at special quantity discounts for business or sales promotional use. For information, please email special_sales@mitpress.mit.edu.

This book was set in Sabon by the MIT Press. Printed and bound in the United States of America.

Library of Congress Cataloging-in-Publication Data

Krasny, Marianne E.
Civic ecology : adaptation and transformation from the ground up / Marianne Krasny and Keith Tidball.
    pages   cm.—(Urban and industrial environments)
Includes bibliographical references and index.
ISBN 978-0-262-02865-3 (hardcover : alk. paper) 1. Urban ecology (Sociology)
2. Human ecology. 3. Community development.  4. Environmental protection—Citizen participation. I. Tidball, Keith G. II. Title.
HT241.K73 2015
307.76—dc23
2014021823

10  9  8  7  6  5  4  3  2  1

I dedicate this book to Akiima Price and Nancy Whitmore,
two ladies who are adaptive and even transformative.
Marianne E. Krasny

I dedicate this book to Seth G. Tidball I and his wife Charlotte, who
served as role models of service and civic ecology until their passing.
Keith G. Tidball

# Contents

# Foreword

As the fog lifted and the sun came out one June morning in Los Angeles in 2003, thousands of bike riders and walkers entered the historic Arroyo Seco Parkway, better known as the Pasadena Freeway. This first freeway/highway built in the West has over the years continued to be the deadliest in Southern California, as measured by the number and severity of car accidents. But on this day, the freeway was closed off for cars, allowing the bikers and the walkers to take over an eight-mile stretch of the roadway. It was an extraordinary moment, revealing to participants a different kind of connection to their surroundings and to the nature so often disguised in the city. One resident, who lived adjacent to the freeway, described how liberating—and disorienting—it had been to "open my window in the morning and hear birds and the wind and breathe the air in a way I had never experienced before." A local environmental activist, also surprised at how transformative the event had been, wrote about hearing birds sing and seeing red-tailed hawks, bullock's orioles, and red-shouldered hawks nesting near the roadway.

Several years later, beginning in 2010 and now taking place three times each year, bikers and walkers can once again experience neighborhoods and streets in Los Angeles in a different way. This event, called CicLAvia, involves more than a hundred thousand bike riders and pedestrians who claim long stretches of surface streets to ride and walk through the city. It's another magical moment for a city that has long had the reputation of a place where no one walks (or bikes) but is demonstrating what Marianne Krasny and Keith Tidball would call a "civic ecology" response.

In the same vein, again in Los Angeles, a local battered women's shelter pulled together a program where the shelter's participants could participate in and help nurture a community garden in a secure site. The battered women's shelter garden project, a form of horticultural therapy, a practice that now extends to prisons, hospitals, and other places where growing the food provides a form of healing, also had a transformative effect on many of its participants. To work the soil, as one participant exclaimed, made her feel "as if the earth [was absorbing] negativity." "The garden gives me air," another participant said, who described her experience as representing a "return to life" where she could stop and "notice the flowers."

Both these situations provide wonderful illustrations of Krasny and Tidball's vision of understanding the ways in which "humans are reconnecting to nature under the most difficult of conditions, and, in so doing, are making meaningful and measurable change to their communities and the environment." For Krasny and Tidball and their Civic Ecology Lab at Cornell University, civic ecology entails a sweeping agenda. It includes reconstructing an attachment to place. It refers to making connections to nature, including in the most desolate or deprived communities. It involves creating new types of social-ecological networks, especially responding to natural disasters such as Hurricane Katrina or Hurricane Sandy (unofficially known as "Superstorm Sandy"), or rebuilding areas devastated by financial crises, job loss, or the huge income disparities that characterize so many urban and rural places. Their research represents a form of engaged scholarship, combining analysis with storytelling. And it involves identifying tools for change.

We are therefore pleased to have this important and valuable book—and their community-engaged scholarship—reside in our Urban and Industrial Environments series at MIT Press. It is a restorative and uplifting, radical and practical, critical and hopeful source of information. And it contributes to a literature that informs, challenges, and ultimately brings a message of hope based on our ability to reconstruct the places—whether in the abandoned neighborhoods in Detroit or in re-envisioning the concrete channel that had once been an urban river in Los Angeles—that are so often damaged in our urban and industrial environments.

Robert Gottlieb, Occidental College, Los Angeles
February 2014

# Preface: Civic Ecology: What and Why?

An excavator rumbled across the dirt, its bucket pushing against the charred remains of a burnt-out house. As the blackened walls splintered and collapsed, acrid dust billowed outward. The jagged debris was quickly scooped up and discharged into the dump truck standing by. Having completed the aboveground demolition, the excavator turned to scrape rubble off the site, and poured it crashing down into the shell of the basement. In four hours, the demolition was complete.

An abruptly silent, now barren lot bore the scars of what just two weeks earlier had been a neatly painted home. But the family had moved out, leaving behind an empty house. Within days, the empty house was stripped of everything and anything of worth—copper wiring, appliances, even bricks. Drug dealers and gang members moved in to occupy the stripped-out former home. A couple weeks later, rival gangs set the shell of what had been a home on fire. Within two weeks of the well-kept home being abandoned, demolition crews were scraping up its remains. Driving down the street where the house had stood was like driving through a dystopia frozen in time—intact homes next to boarded-up homes, homes for sale or for rent nearly touching burnt-out shells of homes, the remains of homes burnt so intensely that they had collapsed, abandoned lots littered with debris, and the scars left on the land after everything was carted away.

This was Detroit in the early 2000s. But scenes not so unlike Detroit can be found in other postindustrial U.S. cities, like Baltimore, Cleveland, or even Washington, DC. Neighborhoods of destroyed and abandoned houses also characterized parts of New Orleans after Hurricane Katrina, New York City after Hurricane Sandy, and Joplin, Missouri, after the 2011 tornado. These neighborhoods are the broken places that poverty and crime, storms and flooding, disinvestment and environmental deterioration, and even war are creating in cities across the United States and in other countries at the beginning of the twenty-first century.[1]

What will these broken places look like by the middle of our century? We have no way of knowing. What we do know is that people in Detroit and Joplin, in New Orleans and New York—and from the Cape Flats townships in South Africa to Korean villages abandoned by all but the elderly—are coming together to re-create and restore these once broken places. They are the stewards who rebuild the health of their community while restoring the health of their local environment. And in so doing they are sending an important message. Humans are not separate from the environment that surrounds us. Humans do not act solely as an external, or even evil force to destroy nature. Rather we are part of nature, and in coming together to heal our local environments, we also heal our communities and ourselves. This is the message of civic ecology practices emerging in broken places. And it is the message of this book.

### Why a Book on Civic Ecology?

In 2008, the world was confronted with a new reality—the majority of its people were now living in cities. Just four years later, Hurricane Sandy struck New York, revealing how vulnerable even a powerful metropolis is to a natural disaster, now compounded by climate change and rising sea levels. But this wasn't the first time the seemingly invincible city had been struck—like New Delhi, Nairobi, and other cities around the world, New York had been unable to protect itself from a deadly terrorist attack just twelve years earlier.

Up until the twentieth century, interaction with nature in relatively pristine natural areas was the norm. Now and increasingly in the future, humans interact with nature in cities—often during times of gradual economic, social, and environmental decline as well as after major disasters and even violence. But what models do we have to guide us away from a dystopian future like parts of Detroit in the early 2000s? For humans to move toward a more positive trajectory in the face of linked social and environmental upheaval, we need positive examples of people acting to restore nature and renew communities. Civic ecology practices provide such examples.

## What Is Civic Ecology?

This book is about the people and practices that are transforming broken places like Detroit, New York after 9/11 and following Hurricane Sandy, and the depopulated villages of South Korea. The people—civic ecology stewards—envision places where humans are neither divorced from nature nor separated from each other. They understand how our future depends on understanding that humans—like other living beings—are all part of a larger nature. And armed with this understanding, they seek to restore green spaces as civic spaces—and civic spaces as green spaces. Their actions embody a longing for—or perhaps a collective memory of—nature, place, and community. And this book is also about understanding the broader implications of these actions, an understanding gained through exploring new thinking about humans' relationships with the rest of nature, about stewardship and civic engagement, and about ecosystems and the ways we govern ourselves. A key to that understanding is the concept of resilience—the ability to adapt and transform, as individuals, communities, and ecosystems, in the face of change and even hardship.

Across the United States and in other countries, people—young and old, rich and poor, new immigrants and those deeply rooted in place—are engaging in stewardship of nature and community. They are revitalizing the broken places left by destruction and disinvestment. Often they are working in places like New Orleans—planting live oaks to express that they, like live oak trees, are rooted in communities that others assumed would never come back after Hurricane Katrina. They are the Capuchin monks planting vegetables in barren lots in Detroit, and the volunteers who install artificial oyster reefs to bring back a healthier Hudson River estuary. And they are the elderly refugees who escaped war in the former Yugoslavia and diaspora in Southeast Asia, carrying seeds with them across land and sea to plant in community gardens in Berlin and Sacramento—as well as the apartment dwellers who create gardens that transform asphalt rooftops into small "islands of community up in the sky."[2]

Yet civic ecology is more than thousands of individual practices that restore nature and community in broken places around the world. Civic ecology also entails applying an ecological perspective to understand how people, practices, communities, and the environment interact. Although the term "ecology" is used to describe the interactions among plants, animals, soils, and streams—the biological and physical components of the environment—it also implies other types of connections. Ecology is more than hierarchical notions of who eats whom (as in food chains). Ecologists, working with social scientists, increasingly are looking at feedbacks, such as how people coming together to plant a tree connect with each other, and how such connections make us more likely to work collectively on future greening activities.

We have deliberately chosen the term civic *ecology* to describe the study of community environmental stewardship practices.[3] This is because, in addition to understanding the stewardship practices per se, we seek to understand their feedbacks and other interactions in the systems of which they are a part. These interactions are with the things we can touch like plants, dirt, and gardens, and less tangible factors like crime and poverty, or flooding and disinvestment.[4] An ecological perspective suggests questions about interactions among life and social and ecological processes. Does stewarding nature help build a sense of community? Does getting to know your neighbors make it more likely that you will engage in stewarding nearby open areas? And do crime and poverty, or hurricanes and flooding, provide the stimulus for people to engage in civic ecology practices? Civic ecology is the study of how community environmental stewardship practices interact with the people and other organisms, neighborhoods, governments, nonprofit and business organizations, and ecosystems in which they take place.

### Two Pillars of Civic Ecology

Two historic figures—one who documented civic life in early America and the other who chronicled his family's efforts to restore degraded farmland—shape the way we think about civic engagement and nature stewardship. Historian and political thinker Alexis de Tocqueville left his native France to travel around the United States during the 1830s. He recorded early forms of civic participation—people in small towns actively engaged in local voluntary associations and local governing.[5] As the nation grew, so did civil society—local civic organizations such as farmers' granges, gardening clubs, and sewing bees dotted the rural landscape. However, as the nation became industrialized and urbanized, many of these traditional manifestations of civil society changed or died off completely.[6] Although some have decried the decline of civic engagement in the United States, others have asked: Is civic engagement not really dead, but rather are new forms of engagement—such as those that involve nature stewardship—replacing the old?[7]

Nearly a hundred years after Tocqueville's musings about civil society in early America, conservationist and philosopher Aldo Leopold wrote *Sand County Almanac.*[8] Leopold's book described his work and observations in rural Wisconsin—and brought about fundamental changes in the way we think about humans' relationship to nature. Perhaps he helped reset our moral compass, which during several hundred years of industrialization had veered dangerously into a mindset where humans no longer considered themselves part of nature.

Leopold was the first professor of wildlife management in the United States. But he was not afraid to leave behind the protected halls of the University of Wisconsin to partake in the hard physical labor of restoring despoiled land. He worked with his family to bring back to health an abandoned farm whose soils had been depleted by years of abuse. Reflecting on the meaning of his labor, Leopold described a land ethic that "changes the role of *Homo sapiens* from conqueror of the land-community to plain member and citizen of it." Being a member and citizen "implies respect for his fellow-members, and also respect for the community as such." And Leopold called for enlarging the "boundaries of our community to include soils, waters, plants, and animals, or collectively: the land."[9] Civic ecology practices heed Leopold's writings about humans' relationship with the nature and community that surround us—and they are deeply rooted in land and in place.

In short, we link the two pillars—civic and ecology—to reflect the bonds between traditions of engagement in civil society and of a land ethic based on humans' deep connections to nature. Civic ecologists realize these traditions through engaging in hands-on restoration and ongoing stewardship of land, life, and community. Through such caring actions, civic ecology stewards develop an ongoing relationship with the rest of nature that contributes to their own and their community's well-being, and even survival.

## Other Uses of the Term

We are not the first to use the term "civic ecology." Among the most notable users of the phrase are scholars writing about civic engagement, and architects and planners attempting to integrate ecosystem processes into community planning and infrastructure.

• University of Wisconsin Professor Lewis Friedland uses the term "civic ecology" to refer to the businesses, nonprofit organizations, and government agencies that are the building blocks for community problem solving and that determine how young people are socialized into civic life.[10]

• For architect and planner Timothy Smith, civic ecology is a "community software" approach to sustainability. It includes local food production, social networks, community governance, and integrated land use and transportation. Together these software components contribute to the community's future "hardware," that is, green buildings and sustainable infrastructure.[11]

• Retaining the focus on urban planning but venturing more deeply into notions of ecology and nature, landscape architect Kathy Poole writes about civic ecology as "a civic realm rooted in ecological sensibility." She describes how "the 'territory' of the city may be extended to consider the natural systems that sustain the city. The hydrology, the geomorphology, the vegetation cycles that comprise the city may be regarded as significant civic structures, worthy of expression."[12]

• Finally, for Seattle-based landscape architect Kathleen Wolf, civic ecology is about how "people in cities and communities benefit from being involved in environmental projects, how urban ecosystems benefit communities, and how to encourage conservation behavior."[13] Our own venture into civic ecology closely parallels that of Wolf and her U.S. Forest Service colleagues working in cities.[14]

### Other "Ecologies"

At Cornell University, we use the term "civic ecology" to refer not only to a diverse set of community environmental stewardship practices, but also to a framework for understanding the role these practices play in larger social-ecological systems.[15] Whereas the impacts of any one civic ecology practice are small and localized, together the feedbacks and networks among myriad practices have important outcomes for individuals, communities, and ecosystems.

Other scholars have proposed similar terms for somewhat different linked social and ecological processes.

- *Human ecology* encompasses a broad set of ideas about human-environment interactions.[16]
- *Public ecology* attempts to integrate not only natural and social processes but also the different types of knowledge held by public-minded scholars, policymakers, and citizens, with the goal of creating more equitable public land and water management policies.[17]
- *Social ecology* captures a diverse set of meanings:

    o a radical stance against political domination, hierarchy, and capitalism as causes of environmental degradation;[18]

    o environmental justice, including advocacy and rights to a clean environment;[19]

    o participatory approaches involving private citizens, government, and the nonprofit sector in conservation planning;[20]

    o research that investigates processes in social-ecological systems;[21] and

    o the series of "nested" systems within which children grow up. These human systems begin with the family, which is embedded in a larger system of school, then neighborhood, and finally community.[22]

- Last, we come to *nested ecology*, which, according to university professor and Presbyterian minister Edward Wimberley, helps us understand the "place of humans in the ecological hierarchy." Wimberley expands on the notion of smaller systems nested in larger ones, as he unfolds a personal or individual ecology, which is embedded in a social ecology, which in turn is nested in an environmental ecology on up through a cosmic ecology and the ecology of the unknown.[23]

## Civic Ecology: Three Distinctions

As distinguished from other "ecologies," civic ecology focuses in on hands-on stewardship practices that integrate civic and environmental values. It also focuses outward on the interactions of these practices with other components of a social-ecological system, including green infrastructure and disturbance. In so doing, civic ecology differs from the ecologies noted earlier in three ways.

First, civic ecology shifts the focus from power, individual rights, advocacy, and planning to *people acting as stewards* of their environment and the community through such practices as allotment gardening, tree planting, and restoring watersheds. These practices create opportunities for individuals in cities and elsewhere to experience the health, restorative, and social benefits of community environmental stewardship, and to learn through their interactions with fellow stewards and with the environment. Furthermore, through processes known as feedbacks, the well-being and learning outcomes of civic ecology practices in turn expand these practices beyond individuals to families, neighborhoods, cities, and potentially, whole societies.

Second, civic ecology employs a *social-ecological systems framework*, including recent work in governance[24] and resilience,[25] to examine the interactions of stewardship practices with organizations, organisms, and the physical environment. Although awkward, the term "social-ecological" reflects the difficulty in trying to separate something "natural" from that which is built or created by humans. Did atmospheric warming caused by humans interact with "natural" atmospheric processes to cause Hurricane Sandy? Or when a business moves out of Detroit, does it leave behind a vacant lot or a "natural" area? Because social and ecological processes are so tightly intertwined, scientists use the term "social-ecological systems" when talking about cities, agricultural lands, a fishing village, and even mountainous regions.

Third, civic ecology addresses head on a fundamental misperception—that is, that humans are exempt from nature. Perhaps the industrial revolution and modern science are to blame, but somehow, over the last two centuries, our species seems to have forgotten our evolutionary past. A self-imposed "exile" from other forms of life has led us to forget how humans lived as predators and prey, nurturers of and nurtured by, and vitally connected to, the living environment that surrounds. Civic ecology fundamentally is an effort to remember—to *put the human and nonhuman pieces back together* so as to nurture places that have become broken due to our forgetting. Our hope is that civic ecology will reframe our understanding not only of the capacity of people in cities and after disasters to restore and revitalize broken places, but also of their tenacious and unfaltering will to do so. Our hope also is that civic ecology will help reframe our fundamental understanding of ourselves in relation to the ecosystems within which we came to be, upon which we depend, and of which we continue to be a part.

**Much in Common with . . .**

Finally, civic ecology is distinct from, but has much in common with the following endeavors.

• *Civic environmentalism* is the face of a new environmental movement in the United States. It came about when people realized that the old adversarial movement—environmentalists constantly battling big business—was failing to achieve either side's goals. And it is viewed as an alternative to top-down government regulations, which while effective in reducing major "point sources" of pollution from factories and power plants, is ill equipped to address increasingly complex environmental problems.[26] Civic environmentalism brings multiple voices to the table to seek collaborative solutions, and is part of a growing civic renewal movement. At its core, civic renewal is about the citizen's role in the production of public goods like shared green space and healthy rivers. And community gardening, organized friends of parks groups, and other civic ecology practices are one means by which people come together to produce such public goods. In this way, civic ecology *practices* are part of the civic environmental movement, but civic ecology is a field of study rather than a movement.

• Often our colleagues confuse civic ecology with *citizen science*. In citizen science, lay people collect data for scientific research.[27] Although the data may be used to help set conservation policies, as when volunteers collecting information about birds identify trouble spots where avian populations are declining, citizen science does not emphasize engagement in hands-on environmental stewardship.[28] Nor does it include civic ecology's focus on community renewal and resilience.

• Civic ecology also is not be confused with the *environmental justice* movement, which seeks to correct the injustices in environmental policies that lead to siting toxic industries, power plants, and hazardous waste facilities in poor communities and communities of color.[29] Civic ecology practices often emerge in poor and in minority communities plagued by toxics and brownfields, but emphasize civic renewal and greening blighted areas, rather than addressing underlying environmental and social policies.

• Two other related areas are *biophilic design*, which outlines how to design buildings that reflect nature's principles,[30] and *biophilic cities*, which are planned so that people have access to nature.[31] Whereas these endeavors may mention public engagement and hands-on stewardship, the focus is on professional design and planning.

• Civic recreation is closely related to civic ecology practices. The thousands of hikers and bicyclists who reclaim the Los Angeles streets during CicLAvia events[32] constitute one such group of civic recreationalists. These and other civic recreationalists are just one step away from being civic ecology stewards, and in fact civic recreation often serves as a gateway to civic ecology practices. After Hurricane Sandy the surfers who saw their beaches devastated became active in restoring New York's sand dunes,[33] just as veterans who use fishing as a means to reconnect with the landscape help restore salmon streams in California.[34]

• Erika Svendsen, Lindsay Campbell, and their U.S. Forest Service and university associates use the terms "urban ecological stewardship" and "civic environmental stewardship" to describe the work of extensive networks of nonprofit, government, and university groups involved in advocacy, education, environmental stewardship, and other environmental activities in U.S. cities.[35] Whereas some of the organizations that are part of these urban networks engage in hands-on nature stewardship or civic ecology practices, others focus on important work like advocacy and mapping green areas. Civic ecology is closely aligned with this work, but differs in our use of a social-ecological systems framework to understand a more limited set of hands-on, nature stewardship practices—stewardship that entails digging the dirt, planting a tree, measuring an oyster, or picking up a piece of trash.[36]

• Civic ecology practices can play a role in *mitigating and adapting to climate change*. Community gardens that produce foods with minimal input, or trees planted by volunteers that absorb carbon, help reduce our environmental impacts. And bioswale gardens that absorb runoff from parking lots or oyster reefs placed so as to protect our shoreline represent community-based adaptations to intense flooding brought about by climate change. Civic ecology practices can be viewed as community-based efforts to respond to multiple stresses—whether they are a hurricane in

New York or economic disinvestment in Detroit. Although small in scale, they provide the hope and the determination to continue in the face of ever-greater stresses, or possibly even the collapse of local social-ecological systems, brought about by climate change. In this way, civic ecology practices can be one of multiple strategies operating at different levels, from the community to the planet, that are desperately needed to address climate change.

**What's Next?**

In this book, we ask four questions:

• Why do people turn to community environmental stewardship in cities and other harsh environments?

• What are the parts—the communities, the memories, the places, the people, and the ecosystem, health, and learning outcomes—that pieced together become civic ecology practices?

• How do civic ecology practices interact with the systems surrounding them, including governance and larger-scale social-ecological systems?

• How might policymakers benefit from and support civic ecology practices?

Following this preface, we outline ten civic ecology principles; each of the ten principles is the subject of one of the subsequent ten chapters. The ten chapters are organized into four parts: Emergence: Why Do Civic Ecology Practices Happen?; Bricolage: Piecing the Practice Together; Zooming Out: A Systems Perspective; and Policy Entrepreneurs: Understanding and Enabling. Each part corresponds to one of the four questions above. Between parts, we provide an "intermission" or "interlude"—a one-page summary of what went before and what comes next—presented in text and through art. And to break up the longest part (Bricolage), we include a photo gallery—a canvas of civic ecology places. You also will find interspersed among the chapters civic ecology "steward stories"—recounting how unbroken spirits are leading their communities in healing broken places.

In answering the four questions, we apply current thinking about human-nature interactions, social and biological memories, communities and civic renewal, learning, environmental governance, social-ecological systems, and resilience. In so doing, we make accessible to the reader the wealth of intriguing conceptual frameworks emerging from scientists working at the intersection of ecology and the social sciences. Having made our arguments for how civic ecology practices present possibilities for addressing concerns about environment and community in a society subject to an increasing number of social and environmental stresses, we turn in the final chapter to how policymakers can act to reinforce these efforts and their positive outcomes.

In attempting to answer our questions, we as authors at times stumble with terms to express, explain, and advocate for what we have observed, feel, and believe in. We recognize that "the whole idea of nature as something separate from human existence is a lie." And that "Humans and nature construct one another."[37] Herein lays the paradox and the challenge. On one level, the use of the term "nature" is fraught with difficulties because it implies that humans and nature are separate. But on another level, we understand intuitively that we interact with and change nature, and that nature has meaning for our human lives—like when we contemplate mist rising from a pond at dawn, or when a child screams with excitement and fear upon encountering a praying mantis ambushing a moth along a garden pathway. These contradictions reflect much of what civic ecology is about. It is about realizing that humans are part of rather than separate from nature, but it is also about realizing the healing and restorative power of nature and of stewarding nature. While recognizing the contradictions inherent in the word "nature," we use the term as is common practice to refer to living things and landscapes, like insects, trees, and green areas, including in cities.[38]

Further, prior to answering our questions throughout the chapters of this book, we acknowledge our passions, our strengths, and our limitations. We have participated in and initiated civic ecology practices. And we are fascinated and inspired by others who have stood up in agonizingly difficult times to recognize the power of greening and community. Following our passion for the people and practices of civic ecology, we pursue a tradition of "appreciative inquiry." Thus at times we may be guilty of looking at the positive side of things—that is the positive aspects of civic ecology practices—and of lacking a critical and comprehensive perspective.[39] At the same time, we recognize that civic ecology practices can be critiqued as an attempt to glorify instances where government abdicates its responsibilities to citizens, leaving unpaid volunteers to do the important work of conserving environments and revitalizing communities.[40] We also recognize that civic ecology practices may contribute to neighborhood gentrification, and eventually to the forced exodus of the low-income people who initially organized and worked hard to green their neighborhoods. We thus commend the work of those who address important issues related to economic class, property rights, and power relations.[41] A whole other area we acknowledge but don't focus on is factors that might limit the emergence of civic ecology practices. Disturbances that fail to reach a certain threshold, governments that limit civil society, and societies lacking a tradition of civic volunteerism may explain the absence of civic ecology practices in some settings.[42] Further, we make claims about civic ecology practices for which we have only indirect evidence from studies of related practices (which are cited in the text). Purposefully, parts of the book are speculative; rather than demonstrating definitive proofs, they are meant to open up our thinking about what is already happening and what could be.

Yet our approach matches our three goals:

• To document how people are defying the harsh realities of poor and blighted cities, wartime, and catastrophic disasters to create places that nurture themselves, their communities, and their environment.

• To help the reader understand new thinking that connects the social and ecological sciences, and how this thinking can be applied to real-life practices.

• And to lay the groundwork for future research and policies that leverage civic ecology practices and their outcomes, so that we may better address future disturbance and catastrophic change.

We start on our journey to achieve these goals with the next "principled chapter."

# Acknowledgments

We thank Robert Gottlieb, editor of the MIT Press series on Urban and Industrial Environments, who coordinated the review of our manuscript and wrote the foreword to this book. Particular thanks go to Louise Chawla, who read the entire manuscript and provided detailed feedback, as well as Liz Holmes, Nancy Whitmore, and Eileen Bach for reading earlier versions and offering suggestions and support. We also thank Miranda Martin at MIT Press for answering numerous questions throughout the process of writing the book. We are particularly grateful for the colleagues who contributed to our thinking and researching for the stories that have gone into this book. They include Alex Kudryavtsev, who lived and contributed to civic ecology practices in the Bronx for one and a half years while conducting his PhD research, and has been an unflagging source of enthusiasm and support for the work of civic ecology for ten years. Richard Stedman always was willing to take time from his busy schedule to answer questions about sense of place, topophilia, and social capital, and to engage in endless discussions about biophilia, red zones, people, and stewardship. Others who contributed ideas and overall support include Betty Baer, Bryce DuBois, Zahra Golshani, Hiromi Kobori, Eunju Lee, Akiima Price, and Mark Whitmore.

Finally, and most important, we thank the civic ecology stewards who have inspired us along the way, including Dennis Chestnut, Jean Fahr, Helga Garduhn, Ben Haberthur, Robert Hughes, the late Theo Manuel, Nilka Martell, Mandla Mentoor, Kazem Nadjariun, Nam-Sun Park, Monique Pilié, Marian Przybilla, J. A. Reynolds, Andre Rivera, and Rebecca Salminen Witt, as well as many young people and adults in the Bronx, Detroit, and New Orleans. We also thank numerous other civic ecology leaders and volunteers from across the United States and around the globe.

# The Principled Chapter

No one more than the residents of this area know what magic has been wrought here at Watts Branch.
—Lady Bird Johnson, wife of U.S. President Lyndon Baines Johnson, on the occasion of the rededication of Watts Branch Park in Washington, DC

People in the Greater Deanwood community near the Anacostia River still remember the day when the President's wife rode a few short miles from the White House to their neighborhood. It was May 18, 1966, when Lady Bird Johnson, shiny new shovel in hand, proclaimed the revitalization of Watts Branch Park in the heart of Washington, DC. Newly planted flowers, neatly mown grass, and a cleaned-up stream bank welcomed the First Lady and her neighbors—the DC residents, displaced by highway building and urban renewal, who were now living in public housing near the park.[1]

Yet, just four years after the First Lady's "magic" pronouncement, funding for park maintenance had dried up, and the magic had turned tragic. For the next thirty years, the site, referred to by locals as "Needle Park," was home to an open-air heroin market, illegal dumping, and rampant crime. The park and surrounding neighborhood had become, once again, a broken place.

But within broken places, unbroken spirits emerge and take action. In the Anacostia neighborhood bordering the Watts Branch tributary, Dennis Chestnut was one of these unbroken spirits. Chestnut grew up in the neighborhood and remembers swimming and wading in Watts Branch on sweltering summer nights with his friends, all the way up to the Kenilworth Landfill. In the creek, the young boys encountered frogs, fish, salamanders, and snakes, but once they reached the landfill, they found a "treasure trove of discovery." Crouching in the bushes to avoid detection by the landfill guards, the boys were on the lookout for the truck delivering three-gallon tubs of discarded Briggs ice cream, too old to sell but not too old to savor. Once the delivery arrived, the boys would jump out from their den and indulge in a "true ice cream Sunday," until the guards caught on and chased them away. Chestnut also remembers the boy who was burned to death in a trash fire at the landfill. He remembers Marvin Gaye as well, who grew up in the Watts Branch neighborhood, and used to sing a cappella with his buddies on the street. But Chestnut would like to forget the generation when pollution fouled the water, trash filled the creek, and junkies roamed the neighborhood.[2]

In the late 1990s, Chestnut was tired of trash strewn in front of his house. He took a broom and decided to sweep up the litter by his house and then, with his kids, sweep up the litter in front of all the houses on his block. A lady on the next block asked him to clean up in front of her house. He said, sure—he would help her if she organized the clean-ups for her block. These fledging efforts grew—in 2001, Chestnut joined forces with twenty-four thousand other DC residents to defy, reclaim, re-create, and redefine Needle Park. They removed over two million pounds of trash, six thousand hypodermic needles, and seventy-five abandoned cars. They helped plant a thousand native trees along the creek corridor. And they gave the park a new name—Marvin Gaye Park—to honor and remember the local boy turned Motown superstar, who sang about the environment, and who died tragically by gunfire at the age of forty-five.

The efforts of Chestnut and others at the park have brought people together around a new sense of community and well-being. Today, grandfathers bring their children to Marvin Gaye Park to run unfettered down its streamside paths, alongside teenage skateboarders and middle-age women walking for exercise. The park provides ecosystem services—like habitat for fish that swim in its cleaner waters, and a green space to absorb contaminated street runoff that otherwise would flow unfiltered into Watts Branch, eventually making its way to the Anacostia and then the Potomac River, and finally to the Chesapeake Bay.

Similar to how the grassy and shrubby areas of Marvin Gaye Park filter out contaminants from the rainwater running into the creek, a Bandalong litter trap waylays the plastic bottles, running shoes, Styrofoam cups, and beat-up soccer balls carried by the creek's current. Young adults, who have lived their lives in DC neighborhoods battered by bullets, banter good-naturedly as they hike up their bright yellow waders and venture into the creek to remove the trapped trash. Through weighing the different types of trash captured in the Bandalong weir, the youth have documented how DC's ban on plastic bags reduced one source of floatables, and thus have helped to shape the city's waste management policies.

The young adults are part of the nonprofit Groundwork's Green Team, as well as DC's Earth Conservation Corps. Since its founding in 1992, nineteen of the Earth Conservation Corps' members have been killed, all but one by drug violence. Yet despite—or perhaps in part because of— witnessing their friends and family fall, the Earth Conservation Corps members have remained defiant. They are determined not only to clean up Watts Branch and Marvin Gaye Park, but also to bring back the bald eagle to Anacostia. The Corps members helped to release Tink, Bennie, and Darrell—young eagles named after their fallen comrades who were trapped by the city's violence—into the wild. Today, the freed eagles nest atop eighty-foot-high oak trees and their fledglings take flight above the Anacostia River.[3]

Earth Conservation Corps is just one partner in a broad coalition of over twenty-five nonprofits, federal and city agencies, and businesses that have come together to take back Watts Branch, the Anacostia River, and the surrounding neighborhoods. Dennis Chestnut, retired from his career as a master carpenter, is now the director of Groundwork-Anacostia, which is the lead community partner in the coalition. Starting with a community's efforts to take back a small park, the coalition has become part of a national network focused on revitalizing rivers and communities.[4]

Through its ups and downs—transforming from a place scarred by crime and pollution, to a place that stakes a claim to healthy recreation, community well-being, and cultural renaissance—Watts Branch has been part of vicious and virtuous cycles, and of cycles of growth, collapse, chaos, and renewal. Come to Marvin Gaye Park on a rainy or sunny day—you might enjoy the sound of running water but be shocked by the volume of trash jammed up behind the Bandalong trap; you can join in the laughter, jibing, and hard work of the young people wading into the creek to clear out the weir, but may be saddened—and inspired—by their life stories. Here you will gain an understanding of what civic ecology is all about.

The story of Dennis Chestnut, the Earth Conservation Corps, the Anacostia River, and its Watts Branch tributary embodies the ten civic ecology principles. In this chapter, we introduce these principles. We start with

Youth monitor and remove trash from the Bandalong trap along the Watts Branch tributary of the Anacostia River, Washington, DC. Photo credit: Alex Russ, Civic Ecology Lab, Cornell University

two principles that explain how civic ecology practices emerge, and then move to five principles that hone in[5] on the parts of the practice—the things that pieced together make a civic ecology practice. Then we zoom out to two principles that place civic ecology practices within larger systems of government, business, and nonprofit organizations, and within even more encompassing social-ecological systems. We end with the final "policy principle." You can explore each of the ten principles in greater depth through reading the next ten chapters, numbered accordingly. A collection of steward stories, a photo gallery of civic ecology places, and three short interludes interspersed among the chapters give life to, and sum up, the ten principles of civic ecology.[6]

## The Ten Principles of Civic Ecology

### 1. Civic ecology practices emerge in broken places

Although today's news is about Detroit and its infamous bankruptcy, not too many years ago similar stories of disinvestment, drugs, guns, factories transformed into brownfields, and abandoned homes and vacant properties that become neighborhood dumping grounds, were about New

---

### Ten Civic Ecology Principles

*Emergence: Why do civic ecology practices happen?*

1. Civic ecology practices emerge in broken places.

2. Because of their love for life and love for the places they have lost, civic ecology stewards defy, reclaim, and recreate these broken places.

*Bricolage: Piecing the practice together*

3. In recreating place, civic ecology practices recreate community.

4. Civic ecology stewards draw on social-ecological memories to recreate places and communities.

5. Civic ecology practices produce ecosystem services.

6. Civic ecology practices foster well-being.

7. Civic ecology practices provide opportunities for learning.

*Zooming Out: A systems perspective*

8. Civic ecology practices start out as local, small-scale innovations and expand to encompass multiple partnerships.

9. Civic ecology practices are embedded in cycles of chaos and renewal, which in turn are nested in social-ecological systems.

*Policy Entrepreneurs: Understanding and enabling*

10. Policymakers have a role to play in growing civic ecology practices.

---

Orleans, New York, Boston, and Washington, DC—cities that now are making a comeback. We call this gradual slide into decline "slow-burn" disturbances, which—similar to war and sudden catastrophes like hurricanes and floods—wreak havoc to places and communities.[7] These are the broken places scattered across our landscapes. But just as dead trees falling to the ground create openings in the forest—opportunities for new life to take hold—so do broken spaces create openings—voids in the landscape waiting to be filled. Civic ecology practices emerge to fill in these voids—the vacant lots, barren median strips, and neglected streams. Civic ecology practices also emerge to fill the voids of landscapes seemingly wiped clear of nature and civilization by sudden and large catastrophes—like tornadoes in Joplin, Missouri, war in Iraq, terrorists attacks in New York, and typhoons in South Korea.

**2. Because of their love for life and love for the places they have lost, civic ecology stewards defy, reclaim, and recreate these broken places.**
How do we explain why Dennis Chestnut decided to take that step into the void—to defy, reclaim, and recreate the neglected Watts Branch and its stream banks?

The urban landscape designer Frederick Law Olmsted realized the restorative effects of nature in envisioning a series of parks in cities across the United States.[8] Over a hundred years later, Yale University professor of social ecology Stephen Kellert described how "the quality of human existence continues to rely on the richness of its connections with natural diversity."[9] Kellert goes on to claim: "Our inborn affinity for the natural world is, in effect, a birthright that must be cultivated and earned."[10]

In talking about humans' birthright to connect with the natural world, Kellert is describing biophilia—our affinity for or love of nature.[11] Biophilia today is expressed through all sorts of activities in nature—hiking, hunting, fishing, or simply sitting by the side of a stream to daydream and reflect. Biophilia also can be expressed through stewardship, including in cities and other places where nature seemingly has been destroyed. Biophilia—and its relative, urgent biophilia, which comes into play under extreme stress[12]—are conveyed and cultivated through civic ecology practices. Civic ecology practices allow anyone—people living in cities, as well as in more suburban and rural areas, and in the aftermath of war and disaster—to reclaim their "biophilic birthright." Perhaps a need to reclaim his biophilic birthright is what led Dennis Chestnut to restore Watts Branch and Marvin Gaye Park.

Chestnut also may have been motivated by his love of a place that was lost. He watched as the creek that had provided opportunities for adventure and exploring nature became contaminated, its channel clogged by trash, and its neighborhood infested with drugs. In addition to biophilia, love of a particular place—or topophilia—compels people to take stewardship action.[13]

**3. In recreating place, civic ecology practices recreate community.**
The young adults who belong to DC's Earth Conservation Corps have created their own close-knit community—an alternative to the gangs that have taken the lives of their friends and family. They have learned to trust each other through working together to remove litter from the Bandalong

traps and from the paths winding their way along Watts Branch creek, through clearing brush and planting native trees, and through raising and releasing bald eagles and other raptors. In a broken place marked by violence, graffiti, and disrespect, it is these small actions—members of the community coming together to demonstrate a collective caring and efficacy—that can spark a cycle of rebuilding and renewal. These small collective acts include civic ecology practices.

**4. Civic ecology stewards draw on social-ecological memories to recreate places and communities.**
Dennis Chestnut's memories of swimming in Watts Branch as a youngster are undoubtedly shared with the other friends he grew up with. These are social memories—the memories held in common by a community.[14] The bald eagles, extirpated from Washington, DC, but with remnant populations surviving in the western United States and Canada, are another sort of memory. They are the biological memories—the seeds, the wildlife, and the ecological processes like decomposition, that are maintained in an ecosystem and that determine its future possibilities.[15] Social-ecological memories are the combination of a community's memories and the ecosystem's memories,[16] and they make civic ecology practices possible. If Chestnut and his Anacostia neighbors no longer had memories of a cleaner Watts Branch, and if seeds of trees native to Washington, DC, were no longer available, then we would not have seen the revitalization of Marvin Gaye Park. And if no surviving bald eagles remained in the world, and knowledge of how they build nests had been lost, then the reintroduction of the bald eagle would have been prevented. When social and biological memories are depleted, social-ecological systems have fewer resources to draw from.[17]

**5. Civic ecology practices produce ecosystem services.**
Ecosystems provide lots of living things that humans need to thrive and survive—insects that pollinate food crops, bacteria that break down human wastes, fish that we eat, and forests that are sites for religious and recreational experiences. Natural areas—the watersheds, forests, and oceans less dominated by humans, and the biodiversity that they harbor—are the source of ecosystem services. Thus, we seek to protect these areas so that they will continue to provide us with essential services.[18]

While recognizing the importance of natural areas and native biodiversity, civic ecology addresses the reality that much of the globe is already occupied by human settlements and cities. In order to maintain ecosystem services, we need to consider a role for those places dominated by humans. Civic ecology stewards restore broken places in cities, and in so doing, create green infrastructure that provides ecosystem services. Further, the civic ecology practices themselves—not just the green spaces they create—provide ecosystem services. Trees planted along Watts Branch help slow the water rushing into the creek during storms, thus providing a water "filtering" ecosystem service. The young adults removing trash from the stream are also providing a filtering ecosystem service—this time filtering out litter and debris. And through affording opportunities for learning and hiking, the cleanups in Marvin Gaye Park provide what are known as cultural ecosystem services.[19]

### 6. Civic ecology practices foster well-being.

In 2006, author Richard Louv sounded the alarm about how children today no longer have the opportunity to swim in a nearby creek like Dennis Chestnut did in the 1950s. Louv's book, *Last Child in the Woods: Saving Our Children from Nature Deficit Disorder,*[20] was a watershed moment in our thinking about the importance of nature to children's health and well-being. It captured the results of hundreds of studies that have shown how spending time in nature can reduce stress, improve physical health, enhance cognition, and even lessen the symptoms of attention deficit disorder. Civic ecology practices encompass spending time in nature, as well as actively stewarding nature, which has additional outcomes for humans. These include a sense of satisfaction that comes from leaving a legacy for the next generation, feeling as if one is able to make a difference, and a sense of empowerment that comes from gaining control over one's life.[21]

### 7. Civic ecology practices provide opportunities for learning.

About a thousand miles north of the Watts Branch, the Rouge River flows through the community of Scarborough, Ontario—a community that has welcomed immigrants from Asia, Africa, the Caribbean, and Latin America. And not unlike what happened in Watts Branch and the Deanwood neighborhood of Anacostia, a civic ecology practice focused on the Rouge River and the Scarborough community has emerged—and provides opportunities for learning. At an event sponsored by the nonprofit Friends of the Rouge Watershed in Scarborough, a local rabbi talked about the meaning of tree planting in the Old Testament, followed by a local imam who spoke on the role of trees in Islam. After the talks, young and old members of the Jewish and Muslim communities joined together to plant trees. Such events introduce participants to other cultures, to working across cultures, and more broadly to the functioning of a diverse and civil society.[22]

What does the interfaith event along the Rouge River in Ontario have to do with the Earth Conservation Corps in Washington, DC—and with learning? Civic ecology practices afford opportunities for learning in uniquely "social-ecological" settings. Through planting trees, civic ecology stewards learn about how to build trust and connections with other people, regardless of background. And they learn about the ways in which trees—and tree planting—contribute to ecosystem services and human well-being. And when they measure which planted trees survive and grow, civic ecology stewards gather important information that they can use to improve and adapt their practices as conditions on the ground change. But not only do civic ecology practices afford multiple opportunities for learning, they also demonstrate how people learn through interacting with each other and with the environment that surrounds them.

**8. Civic ecology practices start out as local, small-scale innovations and expand to encompass multiple partnerships.**
Over time, what begins as a small, self-organized initiative—like Dennis Chestnut joining with the nonprofit Parks and People to transform "Needle Park"—expands into something bigger. The original activists form partnerships with more formal institutions such as local and state government, nonprofit organizations, businesses, and universities. Today Groundwork-Anacostia and Earth Conservation Corps are part of a national network called the Urban Waters Federal Partnership that includes multiple government, business, and nonprofit partners.[23]

Greening practices that emerge after disaster or conflict sometimes begin as small, isolated initiatives and grow to form a network of hundreds or even thousands of partners. For example, after the 9/11 terrorist attacks, people in New York; Shanksville, Pennsylvania; Arlington, Virginia; and Washington, DC, began planting flowers in median strips, parks, yards, and community gardens, and along sidewalks and waterfronts. On

Staten Island alone, organizations like the local chapter of the Federated Garden Clubs of New York, the local botanic garden, a group of senior citizens called Turnaround Friends, Inc., and individuals like resident Wendy Pelligrino, planted healing gardens on barren traffic islands, created wooded walks in existing gardens, and built a new waterfront park. And in a neighborhood in Queens, rendered barren of green by factories, subway tracks, bridges, and highways, students partnered with the Long Island City Roots community garden and the New York Department of Parks and Recreation to plant a single street tree in memory of lost firefighter Michael E. Brennan. The U.S. Congress recognized these community expressions of grief and resilience, and asked the U.S. Forest Service to form a network of support. Thus, out of thousands of spontaneous acts of greening across the country, the Living Memorials Project and network emerged.[24]

**9. Civic ecology practices are embedded in cycles of chaos and renewal, which in turn are nested in social-ecological systems.**
The online Free Dictionary describes a vicious cycle simply as a case where "one trouble leads to another that aggravates the first."[25] Rust Belt cities—where declining employment leads to loss of income and abandonment of neighborhoods, followed by crime and further destruction of properties, loss of jobs, and mass exodus out of the city—are places where one trouble leads to another, each trouble aggravating those before it in a cascading effect. Like vicious cycles, virtuous cycle neighborhoods experience a series of events, each one reinforcing the others, only this time with more positive outcomes. The hope of Dennis Chestnut and his Earth Conservation Corps colleagues is that starting with small acts like cleaning up litter, and moving to more intense actions connecting them with nature and their community like raising and releasing raptors, young people in Anacostia will find alternatives to the vicious cycle of disinvestment, violence, and pollution. Instead the young adults themselves will help to spark a more "virtuous" cycle of civic and environmental renewal, each positive action leading to another.

Vicious cycles help us to understand how neighborhoods become trapped in seemingly hopeless situations, whereas virtuous cycles enable us to envision a different, healthier set of interactions. Now imagine a fifty- or one hundred-year time frame, with periods of crime and disinvestment followed by intervals of community-minded practices and healing neighborhoods. A different kind of cycle—the so-called adaptive cycle—provides a way to understand such cyclic ups and downs over time. To understand the adaptive cycle, visualize a mature stand of trees that is suddenly overwhelmed by a massive forest fire. Immediately after the fire, what was once a verdant, vibrant forest becomes a barren landscape—marred by scorched tree trunks piled at weird angles across the eerie landscape and jutting up into the lingering smoke. This "chaotic" charred landscape is not unlike the barren spaces in parts of Detroit and Washington, DC that have been hit by disinvestment, crime, and arson. But just as seeds sprout, grow into seedlings, and mature into trees that transform the chaos left by the fire into a new forest, people, including civic ecology stewards, start to rebuild their chaotic neighborhoods—which eventually become nurturing places. The adaptive cycle depicts four stages that forests, neighborhoods, and other social-ecological systems cycle through over time—rapid growth, stability, collapse, and rebuilding.[26]

In responding to these ongoing cycles of disturbance, forests, urban neighborhoods, and other social-ecological systems demonstrate resilience. Resilience enables systems to adapt to change, and, most important, to bounce back once they have tipped into decline and chaos.[27] And stewarding nature and community plays a role in the ability of systems to adapt during normal times, and in the ability of chaotic systems to rebound.

**10. Policymakers have a role to play in growing civic ecology practices.**
Several aspects of civic ecology practices, including that they are self-organized through the initiative of local leaders and often take place in cities, present both a challenge and an opportunity for policymakers. The challenge is to recognize and leverage civic ecology practices as existing assets—as social-ecological "innovations" to address environmental and community decline.[28] Recently, we have seen a shift in thinking among policymakers toward favoring such asset-based approaches to community development and sustainability. City governments that may have discounted or even fought against green infrastructure in the 1990s, now actively support the community groups that are planting trees and community gardens, and monitoring and restoring degraded streams.

Asset-based approaches to community development, which seek to identify and build on the resources already present in a community,[29] provide guidance for thinking about policies toward civic ecology practices. Such an approach is reflected in a recent *Bloomberg Businessweek* article, which claims: "The last 50 years have shown that Detroit won't benefit from large-scale actions by the municipal or federal government. Residents have discovered that real recovery comes from community initiatives, entrepreneurial creativity and citizen involvement."[30] This is not to suggest that top-down efforts have no place, and in fact many such efforts, including mayoral sustainability initiatives in cities across the United States, have achieved success in improving local environments and in paving the way for national and international discussions of sustainability.[31] Rather we suggest that homegrown solutions are also important, including in urban communities and in the aftermath of disasters and war, and that recognizing, respecting, and building partnerships with such efforts may contribute to our ability to address what has been called the sustainability crisis. The Environmental Protection Agency (EPA) and other government agencies that formed the Urban Waters Federal Partnership are supporting the efforts of Dennis Chestnut and the Earth Conservation Corps, and bringing these efforts into a national network of like-minded groups operating at different scales. This work of policymakers, who connect practices and bridge across scales, is a step in the right direction.

Thus far we have focused on civic ecology practices along the Anacostia River and its tributaries in Washington, DC, and have taken a peek at other practices like interfaith tree planting in Ontario and memorial gardening after 9/11. In the subsequent chapters, we explore all kinds of civic ecology practices, people, and places—from veterans in Chicago to Nature Cleaners in Tehran, from oyster gardeners in New York to community gardeners in Soweto, and from elders restoring village groves in Korea to teachers and students converting the despised Berlin Wall to a place of greening and reconciliation. We conduct our exploration through the lens of the ten civic ecology principles. And for each principle, we ask: How can the work of scholars venturing into the risky territory of working across traditional academic departmental silos help us to understand the multiple meanings of civic ecology practices?

# I

## Emergence: Why Do Civic Ecology Practices Happen?

# 1
## Broken Places

*Civic ecology practices emerge in broken places.*

Driving down a tree-lined, residential street in Detroit in 2010, one could see all stages of decline: Neatly painted clapboard homes next door to homes that had been abandoned—for sale or for rent signs out front. A boarded-up home and across from that, the charred, still-smoking remains of what had been the home of a retired factory worker. The rubble of homes burnt so intensely that they had collapsed. And interspersed with these houses at successive stages of ruin: vacant lots where all that remained was the imprint on the dirt of what was once a home.

In other parts of the city, this succession from neighborhoods of tree-lined streets and well-kept houses to abandoned land had progressed even further. Areas of Detroit did not resemble what we think of as a city, but rather were reminiscent of a rural, midwestern landscape. Standing in a community garden, one could see a nearby house in the middle of fields overgrown with weeds. A second building—perhaps an abandoned liquor store—was visible off in the distance. Weedy fields that had taken over homes, shops, yards, and sidewalks were now habitat for wildlife—yet remained symbols of lost human habitat. People wandering through an abandoned neighborhood might spook a brood of ring-necked pheasants and watch as they scurried away and awkwardly took flight to land in a nearby tree. Quails, wild turkeys, foxes, and coyotes were now everyday city residents.[1]

The flight of manufacturing jobs from cities like Detroit has been described as a "slow-motion Katrina."[2] It might also be called a "creeping disaster," where the city eventually crosses over a threshold of disinvestment into a very different kind of place. What were once everyday activities, like parents walking their children to school and chatting on the school grounds, disappear as neighborhood schools are shut down, and

Different stages of succession: houses and vacant lots in Detroit, Michigan. Note how the plants are taking over the sidewalk, and how the different patterns of vegetation delineate the outlines of the buildings that previously occupied the lots, as well as the borders of vacant lots themselves, reflecting the different times they were abandoned. Photo credit: Alex S. MacLean, Landslides Aerial Photography

as residents, fearing gang violence, are afraid to walk outside. Along with social interactions, ecological features that residents took for granted, like well-tended neighborhood parks, become fetid dumping grounds. These neighborhoods—the inner cities plagued by disinvestment, gangs, and blight—are reminiscent of places that have been devastated by a sudden, major catastrophe, like a war or tornado. Whether created by a creeping or a sudden disaster, the resulting disfigured neighborhoods become broken places that engender fear and sadness. Yet within these broken places are people whose spirit remains unbroken. And there are abandoned spaces waiting for life to return.

### Crossing Thresholds

Not far from Detroit, in the neighboring state of Wisconsin, lies Lake Mendota. Like Detroit, the landscape around the lake has been changing for years. As it transitioned from forest to farmland, and then to intensive

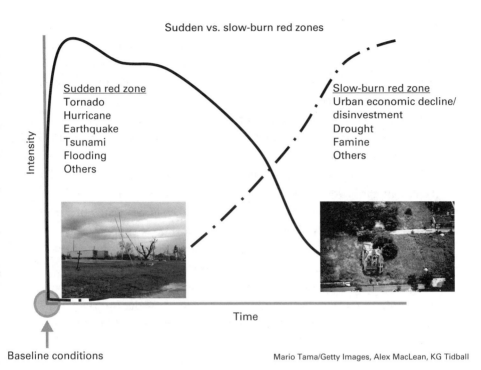

Sudden vs. slow-burn red zones

Intensity

Sudden red zone
Tornado
Hurricane
Earthquake
Tsunami
Flooding
Others

Slow-burn red zone
Urban economic decline/
disinvestment
Drought
Famine
Others

Time

Baseline conditions

Mario Tama/Getty Images, Alex MacLean, KG Tidball

Broken places, also referred to as "red zones," can be created by sudden disturbances or "slow burn" declines. Starting with baseline or "normal" conditions, the impacts of sudden disturbances are intense at first, but decrease over time. In contrast, the intensity of disinvestment and other "slow burn" red zones rises gradually and increases over time. Photo credits: (left) Keith G. Tidball, Civic Ecology Lab, Cornell University; (right) Alex S. MacLean, Landslides Aerial Photography

farming with heavy fertilizer use, more and more nutrients flowed into the lake. Eventually, a threshold[3] was crossed: the phosphorus runoff built up to the point where the lake no longer looked the same, and the organisms and chemicals within the lake no longer "acted" the same. As phosphorus-consuming phytoplankton bloomed, the lake "tipped" from clear to turbid water.[4] Crossing a threshold and entering into a new set of biological, chemical, and social interactions is not new to social-ecological systems: the cod fishery in North America collapsed by the early 1990s after years of overfishing, taking with it the livelihoods of fishing communities in eastern Canada and the northeastern United States,[5] Sixty years earlier, the once-fertile land in the midwestern United States was overexploited by farming, overcome by drought, and transformed into the Dust Bowl. As in the case of Lake Mendota, often conditions seem normal even as scientists and activists are warning of impending disaster. Then, seemingly all of a sudden, a critical threshold is crossed and the entire system collapses. The concern about rising temperatures and climate change is driven by years of observations of these same patterns—a system goes along seemingly as normal, absorbing nutrient runoff, insect attacks, economic variations, and fluctuations in weather—then abruptly, it crosses a threshold.[6]

When Detroit declared bankruptcy in 2013, some people feared that the city might have crossed a threshold. Although unusually wide streets served as a reminder of the Motor City's heyday, large swaths of vacant land had replaced tidy working-class neighborhoods supported by economic prosperity and factory jobs. But Detroit is not unlike other cities and landscapes that historically cycle through periods of growth and stability, followed by economic and environmental stresses, crisis, chaos, and then reorganization.[7] One notorious example is the Bronx in 1977, when a power blackout sparked widespread looting and arson, leading to a near "state of anarchy." By the end of the 1970s, over 97 percent of the buildings in seven Bronx census tracts had been lost to fire and abandonment, and neighborhoods with 50 percent of the buildings razed were common.[8] Similarly, Boston's once thriving Dudley Street neighborhood spiraled downward starting in the 1950s and by 1981, one-third of the land lay vacant. Illegal dumping blighted the area with "auto carcasses, old refrigerators, rotting meat," and construction debris.[9] Cities such as Boston, New York, and Detroit—at times harboring places that are broken and at other times spawning places of growth and rebuilding—span the globe. A satellite image taken at night shows the glow of lights from cities—one can imagine the intensity of the glow varying across the landscape and across time—as the metropolises cycle between growth, collapse, and rebuilding.

### Re-membering

When a fire ravages a forest, it leaves behind a ghostly landscape of blackened snags. Over time, the charred trees fall to the ground. As they decay, the trunk and branches provide nutrients for new plants. And the dead trees leave openings in the forest canopy—opportunities for new life to take hold in patches of sunlight. So too as buildings burn and fall to the ground in ravaged cities, openings inviting new life are created.

Into these abandoned neighborhoods enter the unsung heroes. One such hero, Boston's Dudley Street activist Paul Bothwell, likened the process of rebuilding a community to rebuilding his life after his head and face were crushed when he was hit by a speeding car while riding his bicycle. He talks about the long process of being *"re-membered"* as the broken pieces of his body were picked up and put back together. Rather than an abrupt car crash, the Dudley Street neighborhood suffered a "long process of dismemberment. Literally, it was torn limb from limb and heart from heart and person from person." But just as Bothwell pieced together his shattered body, he describes how "little by little, pieces of this community have sort of re-membered again, pulled back together."[10]

The re-membering process in declining cities sometimes starts with planting a small community garden, inviting in children and adults from the neighborhood, and then expanding to encompass additional elements of community revitalization like retraining ex-prisoners reentering society. Community gardens sprouted up amid the chaos of abandoned neighborhoods throughout New York in the 1970s and 1980s,[11] and were part of the revitalization of the city through the early part of this century.

Then in 2001, terrorists struck the Twin Towers of the World Trade Center. As the towers went down, it seemed as if New York had plunged into chaos once again. For weeks, smoke rising from the piles of debris choked responders and residents. Were the crippled towers symbolic of the new state of the city, if not the country? Had New York reached some sort of threshold, as restaurants and other businesses were closing down, security was heightened, and confusion and fear reigned about the city's future? Despite these doubts and their tragic losses, New Yorkers proved remarkably resilient—ten years later the city once again was thriving, attracting new residents, and experiencing what some saw as a green and civic transformation. Elements of the new, greener New York include public spaces like Brooklyn Bridge Park—rows of derelict warehouses along the East River that have been "re-membered" into a public promenade and restored wetland. And they include the enormously popular (albeit controversial[12]) High Line park. Rising from the remains of a dilapidated elevated rail line that had been slated for demolition, today High Line's suspended pathways landscaped with native plants are one of New York's top tourist attractions, and serve as the nexus of community and economic revitalization in the Lower West Side's former Meatpacking District.

Similarly, in Boston in the 1980s, the Dudley Street Neighborhood Initiative emerged to fill in abandoned lots and voids in municipal leadership, and became the first nonprofit to gain eminent domain over land in a city.[13] Together with community members, nonprofits, businesses, and city government, the Dudley Street Neighborhood Initiative cleaned up toxic sites, provided educational opportunities for young people, and built low-income "green" housing. Six hundred vacant lots were converted to community gardens, where Cape Verdean immigrants and other newcomers to the city grew vegetables from their native islands, re-membering and re-creating their past but in their new homeland.

We see similar civic ecology practices emerging to fill the voids—to re-member the pieces—in Detroit today. The vacant lots that pockmark the urban landscape are braced for new life to return. Plants able to germinate

An abandoned house and adjacent vacant lot transformed into one of many places to pause and reflect on nature's processes, hope, and self-discovery on the Lyndon Greenway in the Cody/Rouge neighborhood of Detroit. The words on the house come from a poem titled "God's Grandeur." G. M. Hopkins, *Poems and Prose*, ed. W. H. Gardner (London: Penguin Classics, 1985). Photo credit: Keith G. Tidball, Civic Ecology Lab, Cornell University

on hard packed surfaces, along with pheasants, raccoons, and rabbits, are some of the historical residents returning to fill the abandoned spaces. They are joined by coalitions of long-term residents and recent arrivals bent on creating something out of the ashes and dirt. They work with the Black Community Food Security Network to cultivate a tree nursery and urban farm in an abandoned city park along the Rouge River.[14] They volunteer with the nonprofit Greening of Detroit to transform large swaths of open space into urban agriculture and nature reserves. And at the Capuchin Soup Kitchen, Catholic monks serve vegetables grown at their Earthworks Urban Farm to hundreds of men, women, and children.

In cities where catastrophic disasters have similarly wiped the landscape clear, civic ecology practices also organize to help fill voids. In New Orleans, a new greening group Hike for KaTREEna emerged after the 2006 hurricanes, while established nonprofits like Parkway Partners transformed their mission to fill the spaces left after the devastation. With the support of these organizations, residents planted young saplings and cared for the damaged, yet still majestic live oak trees that have come to symbolize resilience in their city.[15] In short, regardless of whether vacant spaces are created by slow decline or abrupt catastrophes, groups of people emerge to fill them not just with plants, but also with new forms of action—civic ecology practices that demonstrate the resilience of communities and ecosystems in the wake of decline and devastating loss.

## Unbroken Spirits

Similar to cities and ecosystems, people experience periods of growth and stability, followed by periods of challenge and crisis, in which previously happier and healthier patterns of life are disrupted. When the challenges and crises are extreme, people too may cross a threshold—they break down, abandoning habits and life patterns. But some individuals are able to rebuild their lives quickly, or are able to persist through difficult periods.[16] And sometimes gardening, planting trees, and other forms of cultivating life are part of the rebuilding and resilience process.[17] When that cultivation includes the community, as in civic ecology practices, it contributes to filling the voids in places like Detroit, Boston, New York, and New Orleans, or as we shall see in subsequent chapters, in Cape Town, South Africa, and Tehran, Iran.

The story of J. A. Reynolds and Bruce's Garden in Isham Park, New York, illustrates how one individual—through a desire to improve his community and through a will to overcome personal loss—contributed to filling neglected spaces and voids created by poverty, abandonment, and crime, and by a sudden, unexpected disaster. The story begins when Bruce's family moved to New York in the early 1980s.

When Bruce was 12, gangs wreaked havoc on Isham Park, just down the street from where he lived with his parents. All the benches had been burned. Teenagers drank beer, smoked, cursed.

Bruce's father J. A. Reynolds, a social worker, had a plan. He organized the children into the Park Terrace West Gang, and obtained funding from the New York Department of Youth Services. The "Gang" brought neighborhood youth together to restore the park, and especially the garden.

According to Bruce's father, "We started with a program to corral kids that were misbehaving in the neighborhood. It was pretty bad at the park—lots of drinking, especially. We said 'Look, why don't we take this part of the park and restore it?' We ran a summer program there. . . . Little by little, we cleared the rubble. Then we began planting. Once they were organized and directed, like any group of kids, they responded beautifully. We set up a high school dropout program in that garden and retired neighbors would come out and teach and help the youth with their classes to get their G.E.D.s. Bruce loved working in that garden and never stopped."

Sadly, not only did Reynolds face the inequities of life in New York's gang-stricken neighborhoods. Twenty years later, he faced another, more profound tragedy.

The Bruce Reynolds Memorial Garden in Isham Park sits in Park Terrace, at the northern end of Manhattan. It is dedicated to NYC Police Officer Bruce Reynolds. Officer Reynolds was stationed on the George Washington Bridge when the planes burst into the World Trade Center on September 11, 2001. Along with other members of his police brigade he rushed to the site to assist with the emergency response. He was last seen helping a woman who had been badly burned by jet fuel.

Shortly after Bruce died, a mesh bag of daffodil bulbs, part of a shipment of one million bulbs sent to New York City residents by a Dutch businessman after 9/11, sat untouched under a tree at the corner of J. A. Reynolds's block. "I'm surprised that I am able to sit in the apartment all day and not be bored, just melancholy," Mr. Reynolds said. "I've been avoiding people. When I go out, naturally, everybody wants to talk about him." Finally, the thought dawned on Mr. Reynolds that if he didn't act, his portion of the bulbs would rot. He headed for the garden he and Bruce started so long ago, father, son, and friends, back when they set out to grow good things in a good place.

Through the steady misting drizzle, he toted his pick, his shovel and his trowel. He has no son to bury, but he has flowers to plant.[18]

Reynolds's courageous effort to step outside his apartment and plant daffodils, to memorialize his profound personal loss, was his way of recreating something of value in his own devastated landscape. He was accompanied on his journey by many other New Yorkers and by Americans from across the United States, who, aided by friends from around the world, planted flowers and trees to memorialize the victims of the terrorist attacks. These individuals rejected the idea that 9/11 had become a threshold into chaos for themselves, their communities, or their country. They were defiant and strong—refusing to become idle, apathetic, or broken by tragedy. Instead they converted small sections of gardens, parks, and even highway median strips into living remembrances—symbols of resilience and rebuilding. And in so doing, many drew on their previous experiences with greening, and through renewed greening, helped themselves and their communities to heal.

Reynolds is one of many unsung heroes in the revitalization and resilience of cities. He is one of many who defy horrific conditions, and who demonstrate resilience in the face of devastating losses, through civic ecology practices. What compels these resilient individuals to step into, to re-member, and to heal broken places?

# Steward Story: Monique Pilié

Monique Pilié drove a FedEx truck delivering packages in the Lower 9th Ward of New Orleans before Hurricane Katrina struck.[1] Many of her deliveries were to housebound elderly residents, who depended on her for getting their medications. She came to know the elderly residents on her route, and they came to depend on her not only for their medications, but also for the welcome opportunity to chat when she stopped by with a delivery.

When Pilié heard about the impending storm, she retreated to her family home outside New Orleans. In the midst of the hurricane, she and her family cowered as they listened to tree after tree after tree snap horrifically nearby. One tree crashed down onto their house. She compared the experience to crouching in a World War II foxhole under enemy shelling. For the housebound elderly on her delivery route, the storm was even more horrific. They were trapped when the wall of water breached the dike and swept over their homes. Pilié likens her feelings afterward—as a result of her own experience and her loss of elderly friends who did not survive—to post-traumatic stress disorder.

After the hurricane, Pilié decided to quit her FedEx job and seek the solace of nature as part of her recovery. She embarked on a two-thousand-mile hike of the Appalachian Trail. But she wanted to do more than simply escape the tragedy. For each mile hiked, she pledged to plant one tree in New Orleans—trees that would help the city regain hope and recover. Upon returning to New Orleans, Pilié formed her own tree-planting non-profit group called Hike for KaTREEna, and went on to galvanize hundreds of New Orleans residents to plant thousands of trees across the city.

# 2

# Love of Life, Love of Place

*Because of their love for life and love for the places they have lost, civic ecology stewards defy, reclaim, and re-create these broken places.*

A single black-and-white image—a garden flanking a World War I trench with French soldiers standing in the foreground—haunted landscape architect Kenneth Helphand. He desperately wanted an answer: Why did these soldiers plant a garden in the trenches amid a horrific war? Helphand's book *Defiant Gardens*[1] chronicles a journey to answer this question. It is a history of soldiers in the Vietnam, Gulf, and earlier World Wars, of displaced Jews working as forced laborers during the Nazi occupation of the Polish ghettos, of Japanese in internment camps in the United States, and of survivors of the Bosnian conflict. And it is a history of the small patches of green planted and lovingly cultivated by people living under horrific conditions.

Helphand explores "defiant gardens" from conflict zones around the world throughout the chapters of his book. In the final chapter, "Digging Deeper," he unearths two fundamental reasons why people garden when hope has nearly vanished from their lives, two powerful "loves" that drive them to cultivate plants in an effort to cultivate a future: a love of all that is living, and a love of place. This love of life and love of place propel people around the world to garden when their lives are disrupted by war, as well as by disaster, displacement, and more gradual civic and environmental decline. In this chapter, we explore these two loves, digging deep into the answer to Helphand's question: Why is *gardening* used as an act of defiance in the midst of war?[2] We also reveal how these two loves form the foundation of a conservation ethic.

## Love of Life

Helphand describes World War I trench gardens in France, World War II barbed-wire gardens in prisoner of war (POW) camps in Singapore, stone gardens in Japanese internment camps, and lawns planted by American soldiers outside their tents in the deserts of Iraq. The essence of Helphand's narratives—their skeletons when all detail or flesh is removed—bares itself as love of life versus love of death. The psychologist Erich Fromm proposed two contrasting words to capture these two contrasting loves.

Necrophilia, according to Fromm, is a psychological tendency "directed *against* life.[3] "The necrophilous person loves all that does not grow."[4] He worships all forms of death. Pathological leaders such as Hitler and Stalin exhibit extreme forms of necrophilia.

The opposite of necrophilia is biophilia, whose "essence is love of life."[5] Biophilia, as defined by Fromm, is the "tendency to *preserve* life, and to *fight* death. . . . The person who fully loves life is attracted by the process of life and growth in all spheres."[6]

That humans have an innate love of life—an attraction to other living beings—might come as a surprise. All around us we see, hear, and even smell and taste the opposite: not just extremes like war and mass murder, but also insults to our landscapes and the life they harbor. Farmland converted to strip malls. Plastic bottles strewn along sidewalks and dirt paths, plastic bags hanging from wires, plastics of every kind clogging streams already tainted by sewage. Motorcycles spewing black smoke that chokes slum dwellers and sun-seeking tourists alike, their un-muffled engines masking the sounds of the wind, birds, even human conversation. What passes for real food but is processed into paste and squeezed out from tubes. All of these insults are evidence of an attitude: that we humans are meant to dominate nature to satisfy our desires, even if it means destroying nature along the way. And that we exist outside nature, and are exempt from its laws.

But a keen observer of social animals—including humans—believes there is more to our story. Building on Fromm's biophilia, Harvard University biologist E. O. Wilson has written prolifically about humans' innate love of nature. He hypothesizes that humans, subconsciously, seek connections with the rest of life.[7] Referring to biophilia as an innate propensity to focus "on life and lifelike processes," Wilson states: "To an extent still undervalued in philosophy and religion, our existence depends on this propensity, our spirit is woven from it, hopes rise on its currents."[8]

Together with his Yale University colleague Stephen Kellert, Wilson goes beyond Fromm's notion of biophilia rooted in psychology, to explain how biophilia is rooted in our evolutionary past.[9] Like all species, humans depended on other forms of life for their subsistence, and needed to avoid predatory animals and poisonous plants to survive. Thus, humans became acutely aware of the myriad forms of life around them, and awareness led to an attachment to or even a love of that life. In short, biophilia reflects the fact that we evolved with the rest of the biosphere, not separate or exempt from its natural laws.[10]

The notion that our affiliation with nature is part of our evolutionary past has been hotly contested.[11] In particular, it has been critiqued for its economic class and cultural biases, and for what it implies about people with limited opportunities to spend time in nature.[12] Critics ask: does biophilia predetermine that those mired in urban poverty, or those cut off from nature, will always lead less fulfilling and less productive lives?

Reacting to these critiques, proponents of biophilia acknowledge that, while we are born with an *innate* affinity for nature, we also can *learn* to love nature. Children have an innate tendency to care for plants and animals, but this tendency is reinforced through watching and helping elders care for other living beings. In short, we have an innate fascination with, even a love for, other living beings, but our evolutionary biophilia is reinforced—or not—through our experiences growing up and later as adults.[13] These experiences include civic ecology practices in broken places. But prior to exploring biophilia's connections to civic ecology, we first delve briefly into expressions of biophilia in unexpected places.

## Biophilic Encounters Large and Small

Humans need not wander far from home to encounter evidence of biophilia. The environmental philosopher and activist John Livingston recalled an "extraordinarily encompassing" biophilic encounter in his kitchen: "Very late one night, I was occupying a kitchen stool, he another. Everyone else was asleep. Suddenly, out of long, long silence, he stirred and spoke to me. What he said was brief and not susceptible to translation, but something prompted me to put my head to his, to open, to listen, and to accept."[14] Livingston goes on to describe this encounter with "he" as a gate that opens up our connections to the rest of nature.

There is absolutely nothing unusual about this experience. Anyone who has ever loved a nonhuman being knows the extraordinary encompassing sense of unity that is possible, at least occasionally. I am certain that you know the feeling as well as I do. I suggest, however, that rather than an end in itself this experience is merely the beginning—the opening of a gate that leads to the very nature of being, the nature of life experience and of life preservation. All I ask here is that you allow yourself to extend this selfless "identification"—for that, essentially, is what it is—beyond those individual beings that you "know" in the conventional sense. Touch is good, but not essential. . . . Put your head to the head of the tomcat; the tiger is near.[15]

Who would have thought that Livingston's "extraordinary encompassing" experience of unity with nature came from an encounter with a tomcat occupying a kitchen stool?

In the same vein of finding meaning through a profound experience bonding to other life, but with an animal more majestic than Livingston's tomcat, is the narrative of environmental writer Aldo Leopold's personal transformation:

We reached the old wolf in time to watch a fierce green fire dying in her eyes. I realized then, and have known ever since, that there was something new to me in those eyes—something known only to her and to the mountain. I was young then, and full of trigger-itch; I thought that fewer wolves meant more deer, that no wolves would mean hunters' paradise. But after seeing the green fire die, I sensed that neither the wolf nor the mountain agreed with such a view.[16]

In contrast to Leopold's majestic wolf and even more humble than Livingston's tomcat is the story of how environmental philosopher Arne Næss connected with another form of life:

I was looking through an old fashioned microscope at the dramatic meeting of two drops of different chemicals. At that moment, a flea jumped from a lemming that was strolling along the table and landed in the middle of the acid chemicals. To save it was impossible. It took many minutes for the flea to die. Its movements were dreadfully expressive. Naturally, what I felt was a painful sense of compassion and empathy, but the empathy was *not* basic; rather, it was a process of identification: that "I saw myself in the flea." If I had been *alienated* from the flea, not seeing intuitively anything even resembling myself, the death struggle would have left me feeling indifferent. So there must be identification in order for there to be compassion and, among human beings, solidarity.[17]

The stories of Livingston, Leopold, and Næss suggest that biophilic encounters transcend scale—they can be with a cat on a kitchen stool, a wolf in the mountain wilderness, or a flea in a petri dish—thus making real the possibility for biophilic encounters in cities and seemingly barren landscapes. Returning to a wartime narrative, we are poignantly reminded of the power of even tiny vestiges of life, in this case when hope has vanished, and signs of nature banished. This last story comes from the Warsaw ghetto of World War II.

There were no trees in the ghetto. We called our *Oneg Shabbat*[18] an oasis *b'midbar* (in the desert). But there was no green. One little girl, she was dying, told her sister she would like to see a leaf, to hold something green. Her sister went out, under the wall. The children would pick away bricks and go under the wall. If there were a kind Jewish policeman, they would bring in food. This little girl went to the Aryan side, to a park, and picked a little leaf. That was all. She came back through the hole and put it in a glass by her sister's bed. The other little girl lay there sucking her thumb, smiling. And then she died.[19]

## Urgent Biophilia

We also have experienced and observed encounters with nature in unexpected places. We watched how in New York, a young woman radiated a sense of calm as she rowed her boat on the calm waters of the Bronx River, gliding past a giant heap of jagged scrap metal on the shore, with jets roaring overhead as they came in for landing at LaGuardia Airport. And on a steep hillside in Cap-Haïtien, parents planted trees above a school house—not because they believed the trees would keep the hill from sliding and engulfing the school (which the hill eventually did), but as an act of defiance and hope for the future.

Is what we observe in the actions of soldiers and civilians during times of war and trauma, and of countless displaced persons in cities, something different than the love of life people experience on an everyday basis? Is what drives people to greening in trying times simply the basic biophilia described by E. O. Wilson? How does one explain this intense love of life when everyday life is shattered?

During stable times, humans exhibit a "baseline" affinity toward nature—we plant trees and garden, reflect as we take walks, and enjoy the calm of a canoe floating down a river. However, when faced with rampant destruction threatening our very existence and that of the other living beings we love, the will to have nature as part of our lives, and moreover to care for nature, comes to the fore in unexpected ways. Given that we are conditioned through evolution to have an affinity for other life, when such life is threatened we also feel threatened, making our reactions all the more urgent. This is expressed as an intense—even urgent—desire to affiliate with life. Such intense manifestation of affinity for nature during extreme and threatening situations is called *urgent biophilia*.[20]

Wilson's writings about humans, nature, and biophilia—along with the writings of naturalists about powerful encounters with nature, and our own work observing civic ecology practices in cities and after hurricanes, earthquakes, and conflict—provide a partial answer to Helphand's question about why soldiers and civilians living under the most horrific conditions create and care for gardens. Humans have an innate love for nature, and when threatened, we express this love in unexpected, even urgent, acts of gardening. But biophilia—and urgent biophilia—are expressed not just in wartime, and not just by gardening. They are expressed through slum dwellers planting trees on the steep hillsides of Cap-Haïtien, and through New Yorkers restoring sand dunes after Hurricane Sandy washed away their beaches. Wounded warriors returning from wars in Iraq and Afghanistan express urgent biophilia not just by gardening, but also by hunting, fishing, and restoring wildlife and forested habitats.[21] Children finding escape among small patches of green in violent and polluted cities of South Africa,[22] elderly villagers restoring groves of trees after a typhoon in Korea,[23] and Russians turning to dacha gardening during the social and economic upheaval that followed the collapse of the former Soviet Union[24] are all expressing an urgent biophilia.

To capture these numerous acts of greening, and to expand on Helphand's defiant gardens to encompass a greater range of biophilic behaviors and "red zone"[25] conditions, we ask a question similar to but broader than the question asked by Helphand: Why do people living in a multitude of red zone neighborhoods—neighborhoods that have been abandoned by industry, that are plagued by violence and pollution, and that are devastated not only by war but also by hurricanes, earthquakes, tsunamis, and other disasters—cultivate gardens, plant trees, and otherwise restore and steward these broken places? Urgent biophilia provides one answer. But could other factors in addition to urgent biophilia help answer Helphand's question and our related question?

## Topophilia

Not only frogs, toads, and salamanders but also the marsh they inhabited stirred up profound emotions in Livingston as a child.

There was a marshy area where the water spilled shallowly to one side. There, there were toads and frogs and newts. If you lay very quietly in the grass at the water's edge, you could observe them. The longer you looked, the more deeply you were mesmerized . . . possessed. There was no world, whatever, outside that world . . . nothing beyond shimmering light on water, smooth clean muck, green plants, trickling sounds, flickering tadpoles, living, *being*. That was when the pain started.[26]

The emotion Livingston experienced could be explained by biophilia—a connection with life in the marsh. But he also felt an emotional connection to the marsh itself—to the place of the marsh. Similarly to biophilia, topophilia—or love of place—helps answer Helphand's question

The mural in Bechtel Park, which was knocked down during Hurricane Katrina, reflects New Orleans residents' desire to affiliate with other life, or biophilia. When a desire to associate with other life is expressed urgently following hardship or disaster, we refer to it as "urgent biophilia." Photo credit: Keith G. Tidball, Civic Ecology Lab, Cornell University

about why people turn to gardening in the most desperate situations, or more broadly why people garden, plant trees, restore dunes, and otherwise steward nature in places that experience decline and disaster.

Biophilia—love of life—is connected to topophilia—love of places.[27] Biophilia stems from our evolution when ancient humans needed to know about the bounty and dangers of a particular environment—the plants and animals that they might eat or that might eat them. But knowing about the features of a particular place also has been critical to our survival.[28] It is not too great a jump from knowing about, and feeling an affinity for, the plants and animals in our environment, to knowing about and feeling an attachment to the places that shelter these plants and animals. These places can be the marsh where Livingston observed frogs and newts as a child; the farms where Cambodian refugees to the United States once grew beans, taro, peppers, and eggplant; or the New Orleans that nurtures live oak trees.

Whereas biophilia focuses on life, topophilia is about places that include both the living and the nonliving.[29] Alongside the plants, animals, rivers, and gardens that make up a place, neighbors, social relationships, memories, community history, a feeling of pride in one's community, and even landmarks and buildings all factor into topophilia.[30] Thus, topophilia encompasses how we feel about and participate in the communities that are part of the places where we live and recreate. When places are threatened by development, it is those who have a strong love of place— or topophilia—who are more likely to advocate for improving and conserving those places.[31]

### Restorative Topophilia

Just as soldiers and civilians living under horrific conditions during wartime cultivate plants to bring back life, soldiers, refugees and others who have been displaced from their homelands garden, plant trees, and otherwise steward nature to bring back and to re-create familiar places. Soldiers in the Gulf Wars planted grass seeds and even laid down green tarps outside their tents, to re-create the lawns that reminded them of the places they had left behind when they went off to war. And refugees fleeing from the holocaust in Cambodia hid seeds sewn into the seams of their jackets, later to re-create small plots of foot-long beans, bitter melon, and a colorful assortment of hot peppers, gourds, and amaranth in community gardens in the United States.[32] They, like so many people who have been forced to flee war or escape persecution, recreate their homelands through cultivating familiar plants. And in so doing, they express aspects of topophilia. They create an aesthetic of plants and green spaces. And they simultaneously re-root themselves in their new homes and in a treasured past.[33]

Such acts of "restorative topophilia" occur not only when people are forced to leave the places they love, but also when they watch as the places they love are destroyed. After the Iraq War, Marsh Arabs, whose livelihoods had been decimated when Saddam Hussein diverted water away from their ancient desert wetlands, worked to restore the wetlands and their traditional way of life,[34] thus expressing restorative topophilia. Residents who transform vacant swaths of abandoned land into urban farms in the troubled city of Detroit provide another example of restorative topophilia. Unlike urgent biophilia, which focuses on our strong drive to associate with other living things when faced with disaster or war, in restorative topophilia we draw on our memories of what a place once meant to us; we invite neighbors, communities, even whole societies to reimagine and restore a place. Together we demonstrate faith in the connection between people and the places they inhabit, create, and recreate. Whereas urgent biophilia is an immediate reaction to a threat that propels us to interact with life, restorative topophilia channels that reaction into restoring broken places, and restoring hope for the future.

Like a certain kind of biophilia, a special type of topophilia is expressed when nature around us is destroyed, when beloved places become broken or are taken away. Urgent biophilia is a reaction to this threat, driven by our evolutionary selves. It is expressed as retreating into a natural setting for reflection, throwing ourselves into adding houseplants to our home, or planting a tree or a flower in a nearby outdoor setting. When our reaction is expressed as restoring a place that has become broken, it becomes restorative topophilia.[35] And like love of life, love of place helps answer Helphand's question about soldiers gardening in the trenches of World War I. Like urgent biophilia, restorative topophilia explains why people seek to recreate beloved places, places that have been destroyed, and places that they had to leave behind. People living in in the trenches of World War I, in the Polish ghettos of World War II, and in the desert tents of war-torn Iraq felt a strong, even urgent, need to connect with nature. They also experienced a strong desire to recreate aspects of their home, including the green spaces such as the trees and lawns outside their houses or the wetlands that enabled a way of life. Thus, urgent biophilia and restorative topophilia help answer Helphand's and our questions about why people defiantly turn to greening during extreme and difficult times.[36]

## A Conservation Ethic

John Livingston experienced deep pain when the marsh he loved and the animals that occupied it were decimated by development. This profound loss led him to become an environmental writer, a professor of environmental studies, and a founding trustee of The Nature Conservancy of Canada. Aldo Leopold also experienced the degradation of the land as a personal loss, and went on to write *A Sand County Almanac*[37]—launching the conservation movement in North America and defining a lasting conservation ethic. In taking personal action to restore abused land, and in urging others to become land stewards, Livingston and Leopold were expressing their biophilia and topophilia—their connections with life and with place. And they along with their fellow nature writers and philosophers also articulated a conservation or new "environmental virtue" ethic.[38]

Connections with life and with place—and restoring life and place—are the foundations of civic ecology practice. These connections suggest that humans flourish in and through restoring nature, and that we are part of nature and the places where we live. And these practices demonstrate that human beings can "live well in nature,"[39] through practicing a conservation or environmental virtue ethic—embodied by stewardship of our communities including all forms of life.[40]

In words that reflect "living well in nature," the Norwegian philosopher Arne Næss wrote:

We need environmental ethics, but when people feel that they unselfishly give up, or even sacrifice, their self-interests to show love for nature, this is probably, in the long run, a treacherous basis for conservation. Through identification (with nature), they may come to see that their own interests are served through conservation, through genuine self-love, the love of a widened and deepened self.[41]

In the context of civic ecology, Næss believes that a conservation ethic based on "the love of a widened and deepened self" will serve us better than the treacherous path of reminding people of their selfishness. Civic ecology practices provide opportunities to realize our deepened ecological selves[42] in nature, and thus deny the notion that humans act only to destroy other life through warfare and selfish acts. Similarly, they contradict the notion that human welfare is necessarily opposed to the welfare of our environment. Rather, civic ecology practices are motivated by an attraction to life and even joy, and point to a moral compass that strengthens the life-loving side in ourselves.[43] Like Fromm's "biophile," the civic ecology steward "does not dwell in remorse and guilt, which are, after all, only aspects of self-loathing and sadness. He turns quickly to life and attempts to do good."[44]

The historical accounts of Aldo Leopold and Henry David Thoreau reflect these notions of love of life and love of place. Today, their descriptions of living well in nature are finding a new expression in cities and in broken places, as people take it upon themselves to restore nature and community. Like Thoreau at Walden Pond, the civic ecology stewards who do the work of restoration experience personal growth and enrichment through seeking understanding of themselves and nature. Like Leopold, they experience feelings of health and well-being through actively caring for nature. The actions of Leopold and Thoreau and those of civic ecology stewards are a means for expressing our ethical commitments in concert with nature, rather than as opposed to nature.[45]

And in places where warfare, disaster, and disinvestment create unspeakable conditions, attempts to live well in nature—no matter how small—enable people without other freedoms to defy the reality that surrounds them, to find an escape from their daily suffering, and to demonstrate their defiant will to live, to grow, and to survive.[46]

# Steward Story: Nilka Martell

Nilka Martell[1] has lived in the same apartment at the corner of Virginia and Newbold Avenues in the Bronx since she was born. As a young child, Nilka used to visit Puerto Rico for a couple weeks every summer. She would watch her grandmother pick mangos and avocados or kill one of her chickens for the evening meal. And Nilka would gather coffee beans from the bushes in front of her grandmother's house. Back in the Bronx, she would walk along the waterfront in Pelham Bay Park with her father and two sisters, the beach scenery reminiscent of Puerto Rico. Now grown up, her sisters have left the Bronx for greener and warmer places. But Martell has remained, at first forgetting her early childhood experiences in nature, but now remembering as she tries to find her "own niche with nature in the city."

Just before Christmas in 2010, Martell was laid off from her job as a paralegal. It was cold, she was sad, and she couldn't find work. She needed to get out and do something for herself and for her two teenaged kids.

In March, as things got a little bit warmer, Martell coaxed her son Isaias and daughter Lia Lynn to join her at a Bronx Park cleanup. A few weeks later, she dragged herself out of bed, fixed her usual cup of coffee, got her kids off to school, and stared out the kitchen window. Across the street was the same grocery store she had seen out the same apartment window for over thirty years. But inspired by her volunteer work in the park, she saw things differently this time. The lot next to the grocery store looked bad—much worse than the park she had just helped to clean up. Overcome with high weeds, garbage, and dog waste, the lot was an eyesore that said to the kids and adults who walked by: No one cares for this dead-end block on Newbold Avenue.

Martell asked the grocery store owners if she could clean up the lot. They granted permission but assured Martell that her efforts would be in vain. Over the next couple weeks, Martell, Isaias, and Lia Lynn pulled out the weeds and picked up the dog waste. Neighbors would walk by and ask them why they were bothering to clean up a place in the Bronx. According to Martell, "At first, and to be quite frank, I was just looking to kill boredom! But we created a presence, and local kids joined in the effort."

In 2010, the same year Martell got laid off from her job, the city planted ten trees on her dead end street. Martell knew nothing about trees, why they were important or how to care for them, so she enrolled in a tree care workshop hosted by MillionTreesNYC. Inspired by the training, she and her kids adopted the ten trees, coming out once a week to deliver fifteen to twenty gallons of water to each of them. The neighborhood kids, like Martell a year earlier, were looking for something productive to do and were eager to help out. Adult neighbors also took note that Martell and the kids were caring for their block, and donated flowers and plants to replace the weeds. With just $20 remaining from her unemployment check, Martell purchased planters for the donated flowers and plants, and placed them along the street in front of the cleaned-up lot.

As more and more people joined in the block beautification activities, Martell's initial efforts expanded. She secured a $1,000 grant from the Citizens Committee for New York City to install a dog waste system and had the neighborhood youth convince dog owners to use the receptacle. She and her growing group of volunteers next persuaded the city to replace the dilapidated street signs, remove the graffiti, deliver mulch for the Virginia Avenue trees, and clean up a catch basin installed beneath a grate to collect street runoff, but which had caught trash along with the water. And the volunteers constructed tree guards to protect street trees from dog waste, errant strollers, and reckless bikers. The group gave themselves a name—Getting Involved, Virginia Avenue Efforts, Inc., or G.I.V.E.

At G.I.V.E.'s first big event in 2012, more than 125 volunteers showed up. And the event was featured by Mayor Bloomberg's NYC Service initiative as an example of "setting a new standard for how cities can tap the power of their people to tackle their most pressing challenges." Quoted in NYC Service's annual report, Martell says, "It has been awesome to witness the change in dynamics in our neighborhood through beautification projects. . . . Our common goal of taking care of our block has led us to stronger bonds with community members, business owners, and City agencies."[2]

Since then, the volunteers have branched out from their initial Virginia Avenue block beautification efforts. They have taken responsibility for maintaining the trees in the South Bronx's new Concrete Plant and Starlight waterfront parks. And as part of a citywide "park(ing) lot" day, they cordoned off a parking space in front of Concrete Plant Park, and asked passersby to join them in jumping rope, playing games, chalking the sidewalk, and admiring potted flowers. As folks stopped by, Martell and her volunteers let them know that instead of just walking by Concrete Plant Park every day, they might want to walk through it—and enjoy the trees, grass, relics of the Bronx's industrial past, and revitalized waterfront greenway.

Like other civic ecology practices, the work of Martell, her children, and her community organization is not just about greening. As part of their efforts, G.I.V.E. celebrates the Latino, African, and indigenous Taino cultures of Puerto Rico and the Bronx by inviting a *bomba* group to give free drumming and dancing lessons after the day's greening work is done. They also help to empower elderly neighbors, who cannot do the physical cleanup work, but are elated to show off their savory cilantro, pigeon pea, pork, and rice dishes. And Martell takes teens on historical walks of the Bronx conducted by another community leader, the late Morgan Powell, founder of Bronx River Sankofa.[3] The teens and the elders, immigrants from the Dominican Republic, Puerto Rico, and Mexico, learn English and about each other, through caring for trees, planting Mexican sunflowers, and celebrating their outdoors and their cultural traditions.

But Martell is not content to transform just one block on a dead end street. Instead, she is intent on convincing others that they have the power to make change in the South Bronx. And she is succeeding. Her efforts recently spawned a new community group, "Friends of Starlight Park," which joins G.I.V.E. to prove that in the Bronx, people care and have the power to make change.

# The First Interlude

Art: Gretchen Krasny

Thus far, we have defined what civic ecology is and is not. We have explored the broken places where civic ecology practices emerge. And we have answered Helphand's and our question about why people turn to greening under the most difficult of conditions—how cultivating life and restoring places defies devastating realities, and sparks a new ray of hope for a more nourishing future.

Next we narrow in on the practices themselves. What are the "pieces" that woven together become civic ecology practices? What determines what form a practice will take? And what do the practices contribute to the surrounding community and environment?

In the chapters in part II, Bricolage, we pick up and examine five such "pieces." And along the way, we take a break to explore a gallery of civic ecology places.[1]

# II

Bricolage: Piecing the Practice Together

# 3

## Creating Community, Creating Connections

*In recreating place, civic ecology practices recreate community.*

Guatemala City's *basurero* rises twenty stories above an impoverished neighborhood. Men anxiously await the arrival of dump trucks and scour the loads for anything of value—plastics and glass to be recycled, dilapidated furniture that can be reused, rotting food barely edible. It used to be that entire families of Mayan migrants displaced by war in the Guatemalan highlands lived in the dump, surviving off the city's waste. But in 2005, the methane cloud hovering above the landfill ignited and acrid smoke from burning tires, plastic bags, and even dead human and animal remains permeated the surrounding area, spurring regulations so that today only adults are permitted to scavenge the massive trash heap. But three years after the rules were put in place, landslides killed eight adults along with two children working in a steep area of the landfill.[1]

The regulations forced highlands families out of the *basurero*, displacing them once again—this time to the squatter camps perched along the landfill's perimeter. But the families remained connected to the *basurero*—not only by the work of the remaining garbage pickers, but also by their use of its putrid effluent for water.

### Sense of Community: Lost and Found

When people left the highlands, the community ties they had in small rural villages were lost. And their connections to each other were further eroded when they were displaced once again—from their homes inside the dump to its perimeter. Young men, no longer able to eke out a living as garbage pickers, turned to murderous gangs. The gangs enabled the men to create a semblance of a community that cared for them, however dystopic. Just as the dump and squatter camps were a hellish place, so too the gangs destroyed any sense of safety for the rest of the slum dwellers. Like so many other broken places, this is also a place where families and communities are broken.

The nonprofit organization Safe Passage recognized the role that a safe green space could play in helping to rebuild the sense of community Guatemalan highlanders lost when they fled to the city. Safe Passage acquired decommissioned land from the *basurero* to build a preschool and vocational training center. Then they invited landscape architect Daniel Winterbottom and his students from the University of Washington to plan a garden on the land, which would be capped and planted with trees and shrubs to absorb soil toxins and filter air-borne contaminants. The first step in planning the garden was to uncover the dreams of the community members for a safe space. Adults said they needed places "to socialize, talk and spend time together, knowing their children were safe and engaged in activities of their own." They also volunteered that they wanted a "place with abundant vegetation, lots of color (flowers) and wildlife," along with plants that reminded them of their homeland and "areas to grow food so that they could share their skills with their children." Children wanted "places to explore, climb, play and run," and where they would find insects, fruit trees and open lawn, castles and forts. Today, the Children's Garden of Hope Park provides a safe and nurturing place for children to play, run, and learn. And it invites community members into the safe verdant space through offering a gathering area for mothers, for celebrations, and for outdoor recreation, thus recreating a sense of community for the displaced highlanders.[2]

The gang violence in the Guatemala slums is not unlike that found in other cities around the world, particularly where displaced families are congregated in substandard housing. Perhaps the Mayans fleeing their villages were seeking a "Promised Land" of economic opportunity and relief from the violence of the countryside. This was certainly the case when black Americans fled back-breaking work in the cotton fields of the south with pay so low they could scarcely feed their families, and where the threat of a Ku Klux Klan lynching menaced daily. In his account of black American migration to the "Promised Land" of the north, reporter Nicholas Lemann describes the cramped, run-down housing the newcomers encountered upon their arrival in Chicago in the 1940s and 1950s, and how excited they were when the Robert Taylor Homes were built in 1962.[3] Recounting one family's experience moving into their brand new high-rise apartment in building "5135," Lemann writes:

It was a great day. There was a feeling of excitement and of festivity that went along with the inauguration of an impressive building, especially since the accommodations there were better than any of the tenants had ever had. Janitors were there to help everyone with their things. Workmen were grading the area around the building and planting grass. Everything was new and clean. The only odd note that day was that a Baptist church right across the street from 5135 burned down—but that could be taken as a sign of the momentousness of that spot, rather than a bad omen. The Haynes family chose to rejoice in their good fortune in becoming residents of the Robert Taylor Homes. As Ruby's son Larry, who was twelve years old at the time, says, "I thought that was the beautifullest place in the world."[4]

Shortly after Ruby Haynes moved her family of ten into their new home, Larry's older brother had become a gang leader. He and his gang controlled two buildings in the Robert Taylor Homes project. He was part of the violence and drug wars that made life for those seeking the Promised Land unbearable. In 1988, the gang activity led to four shooting deaths. But by that time, Ruby Haynes had moved back to her rural Mississippi birthplace—to the small town she had left thirty-three years earlier to forge a new life in Chicago. And in 2007, the massive Robert Taylor Homes complex, the place that had once seemed the "beautifullest place in the world" to black American migrants seeking the promised land of the north, was demolished by the city of Chicago.[5]

When families fled the highlands for Guatemala City's garbage dump, or the Mississippi cotton fields for Chicago's public housing, they left behind not only small rural places, but also the meanings and emotions associated with life in these towns and villages. Although life was unbearably poor and often scarred by violence, there was a sense of community that connected the residents to each other.

People who are part of a community feel as if they belong and make a difference to others. They create emotional connections through interacting with their neighbors and friends. And they have a faith that their needs will be met through being part of a group. In short, they have a sense of community that provides them with feelings of belonging, of having influence, of emotional connectedness, and of being cared for. Often a sense of community is closely tied to having access to and caring for green space.[6]

Not only displaced people, but also those who stay put lose this sense of community when the place where they live changes. Traditional livelihoods are replaced by intensive agriculture, a neighborhood is divided by a highway, a marshland is paved over for development, and a beach is washed away in a typhoon. When people leave their homeland for the city, or when they suffer the loss of a cherished place and community, they try to reestablish aspects of the places and the communities they left behind—like the gang members seeking a community in Guatemala City or Chicago, or volunteer members of Nature Cleaners in Tehran.

### Creating Community in Iran

Gang membership is one way to recreate community—feelings of belonging and of having influence or efficacy. Civic ecology practices are another path to create meaningful community. Although picking up litter is often belittled as a small, almost useless act, in some places—like a civic ecology practice in Iran—it takes on new meanings.

In Tehran and other Iranian cities, volunteer members of the group Nature Cleaners clean up trash in popular parks and historic sites. Nature Cleaners started when Kazem Nadjariun became alarmed by trash strewn along the shores of a lake where he was vacationing with his family. He and his family picked up the litter and posted photos on Facebook—and news of someone finally taking a stand against littering went viral across Iran. Within sixteen months after Nadjariun's initial photos were posted, 450 litter cleanup events had been held in parks, historic sites, and other high-visibility open areas across Iran. The Nature Cleaners Facebook site, which is used to share events and offer encouragement, connected forty thousand members in 2013, about three thousand of whom had been active in cleanup events.[7]

For the volunteers, Nature Cleaners activities are not just about cleaning up nature. During an event in the city of Toloran, one hundred Nature Cleaners volunteers joined together to make a big pot of soup, and to partake in a meal together after they had completed the cleanup. Following a trash pickup in Karaj, participants hiked to a nearby stream, chatted, and enjoyed each other's company. Facebook posts refer to "connecting with

Nature Cleaners volunteers in Iran picking up litter on a greened median strip, and creating a sense of community through sharing breakfast tea. Photo credit: Zahra Golshani

others," "creating community," and becoming a "family." By being part of a larger movement, Nature Cleaners volunteers know that together they are having a larger impact and doing something for their country more profound than the simple act of picking up litter suggests. By learning social skills, responsibility, and teamwork through their civic ecology practice, they are contributing to a brighter future for Iran.

Through picking up trash and their related social and civic activities, the Nature Cleaners volunteers are developing feelings of belonging, of having influence, and of connectedness—all aspects of community. Humans want to feel part of a group or community. And being part of a community of people who trust each other also makes us more likely to act for the common good—often starting with small acts like picking up trash in a park.

### Social Capital—An Unlikely Means to an End

Oftentimes when a government official wants something fixed in a city or neighborhood, she will direct people to make the change or simply do it herself. The work of the Nature Cleaners demonstrates an alternative to top-down means of bringing about change—the cleanup events were "self-organized" by Nadjariun, his family, and his fellow citizens. Over a hundred years ago, West Virginia School Supervisor Lyda Hanifan similarly demonstrated an understanding of self-organized means for local change. He leveraged this understanding to devise "unlikely means" to achieve his policy ends.[8]

Early in the twentieth century, dwindling "goodwill, fellowship, mutual sympathy and social intercourse" among rural residents was having a negative impact on local schools, according to School Supervisor Hanifan. To focus attention on the problem and its solution, Hanifan invented a new term. He had no way of knowing that a century later, the term he coined—social capital—would be the subject of thousands of books and papers, much less that it would generate over two hundred million hits from a google search. The ideas his term captured still today help explain how people come together to manage a shared, public resource (like a town forest or regional road system) so that everyone derives some benefit. And they offer a counter-narrative to the popular misconception that each person looks out only for himself or herself, and thus that collectively, humans will always create a "tragedy of the commons."[9]

In explaining his new term, Hanifan began by referring to financial capital, or the accumulation of wealth in business affairs. Is there also a wealth, a "capital," that is accumulated in social affairs?

Now, we may easily pass from the business corporation over to the social corporation, the community, and find many points of similarity. The individual is helpless socially, if left entirely to himself. Even the association of the members of one's own family fails to satisfy that desire which every normal individual has of being with his fellows, of being part of a larger group than the family. If he may come into contact with his neighbor, and they with other neighbors, there will be an accumulation of social capital.[10]

Hanifan went on to elaborate on the impacts of his "social capital": "The community as a whole will benefit by the cooperation of all its parts, while the individual will find in his associations the advantages of the help, the sympathy, and the fellowship of his neighbors."[11]

Guided by Hanifan's notion of social capital in the early 1900s, a West Virginia school district organized a series of community gatherings, including "sociables," picnics, concerts, sports events, and an agricultural fair, which brought together people from throughout the area. The social activities allowed people to have contact with others living nearby and to feel as if they belonged to a group. In so doing, the rural residents accumulated social capital. The social capital also benefited the community as a whole—once connected to each other through community gatherings, residents worked together to successfully advocate for road improvements.

Hanifan summarized his notions of social capital and people working collectively toward the common good:

Tell the people what they ought to do, and they will say in effect, "Mind your own business." But help them to discover for themselves what ought to be done and they will not be satisfied until it is done. First the people must get together. Social capital must be accumulated. Then community improvements may begin. The more the people do for themselves the larger will community social capital become, and the greater will be the dividends upon the social investment.[12]

Hanifan's prescription—to help people "discover for themselves what ought to be done"—remains a common theme today in discussions of social capital and of people's ability to organize themselves to improve their communities. Such "self-organization" is also a fundamental tenet of current-day civic ecology practices.

Foreshadowing an issue that continues to plague discussions of social capital today, Hanifan's work in rural West Virginia begs the question: Is social capital a cause or effect of people coming together? Reflecting this chicken-or-egg scenario, he talks not only about how social capital accumulates "the more people do for themselves." He also states: "In community building as in business organization and expansion there must be an accumulation of capital before constructive work can be done."[13] Later on, we reengage with this chicken-or-egg problem—what we call a "conundrum of circularity."

## Working toward the Common Good

Social capital helps us understand how people come together to work toward a public good—like an improved rural road. The public good in the case of civic ecology might be a newly daylighted river flowing through downtown Yonkers,[14] a pond that provides habitat for dragonflies in the densely populated city of Yokohama,[15] or even the protected wildlife in Jerusalem's Gazelle Valley Urban Nature Park.[16]

To help understand how social capital plays a role in creating public goods, we turn to Elinor Ostrom, a political scientist who was awarded the Nobel Prize for her work on how people collectively manage shared resources. Ostrom observed that not every time a community has control of a common resource do they exploit that resource to extinction—the so-called tragedy of the commons. She spent her career studying exceptions to the tragic rule and trying to find explanations for them. She and her fellow scholars demonstrated that communities with higher levels of social capital are more likely to manage a shared resource like a sacred forest or irrigation system for the common good. Ostrom identified three components of social capital that are present when people successfully manage a shared resource: trustworthiness, social networks, and shared adherence to rules that govern our everyday lives. In other words, communities have social capital when people trust and know each other, when there are clear rules about use of the resource, and when governments do not interfere with and even act to facilitate people working together. These are the same communities that are likely to manage common resources so that the community benefits, and thus avoid the tragedy of the commons.[17]

Under what conditions do individuals come to know and trust each other, and behave in ways that generate mutual benefits? In rural West Virginia, the school superintendent helped communities build social capital through social activities, concerts, and an agricultural fair—and they went on to advocate for better public roads. In Tehran and other cities today, civic ecology practices mirror Hanifan's social activities—they bring people together. In coming together to clean up litter around a lake in Iran, savor the smell of fresh basil they have cultivated in a community garden in Houston, or saw down invasive shrubs to allow native prairie to return near Chicago, people develop trust and create bonds, and engage in action for the common good. Further, when other people stop to observe, converse with participants, and even join in civic ecology activities such as a Nature Cleaners litter cleanup in a Tehran park, the bonds extend beyond those directly engaged in the civic ecology practice to encompass the wider community.

Recalling our "conundrum of circularity" (the chicken or the egg problem), civic ecology practices embody the ways in which social capital both enables and results from community environmental stewardship. In communities where social capital is missing, where trust is weak and conflict is rampant, civic ecology practices can enable people to begin to come together and develop social capital. And in situations where social capital is already strong, it provides a basis for people to act collectively when confronted with environmental or other problems. Thus, we might extricate ourselves from our conundrum by thinking about social capital and civic ecology practices in terms of both development and deployment. Social capital develops when people work together in civic ecology practices—and social capital can be deployed to expand existing civic ecology practices. Such circular or feedback processes are actually quite common despite the fact that we like to think linearly about cause and effect. We will revisit these feedback "develop and deploy" processes in other chapters—including the chapter on how learning develops through and is deployed in civic ecology practices.

### Life Is Easier . . .

Nearly eighty-five years after Hanifan's work to build social capital in rural West Virginia, Harvard political scientist Robert Putnam in his popular book *Bowling Alone*[18] again decried the unraveling social fabric—this time of the entire American society. He pointed to the decline in important forms of civic engagement like voting, attending town or school meetings, serving on committees for local organizations, churchgoing, membership in women's clubs and men's fraternal orders, and volunteering with the Boy Scouts, Girl Scouts, or the Red Cross.[19]

Putnam articulated a broad view of social capital. It includes eleven different domains—ranging from how much you trust people of different ethnicities to how often you volunteer and engage in civic life in your community[20] (see table below). He and other researchers have conducted numerous studies that show how communities with higher social capital have less crime, lower dropout rates, and more effective civic institutions.[21] Thus, according to Putnam, "For a variety of reasons, *life is easier in a community blessed with a substantial stock of social capital.*"[22]

**Social capital domains found in civic ecology practices**

| Social capital domains (according to Putnam)[23] | How many of Putnam's social capital domains can you find reflected in civic ecology practices?[24] |
| --- | --- |
| 1. Social trust<br>2. Interracial trust<br>3. Diversity of friendship network<br>4. Civic leadership<br>5. Associational involvement<br>6. Faith-based social engagement<br>7. Equality of civic engagement across the community<br>8. Informal socializing<br>9. Giving and volunteering<br>10. Conventional politics participation<br>11. Protest politics participation | By pulling weeds, preparing soil, and replanting native species in Seattle's city parks, civic ecology stewards not only become involved in a friends of the park civic association, they also learn to trust the person who hands them the shovel or helps them to navigate the wheelbarrow down a precipitous, rutted pathway.<br><br>Time spent under an oak tree on a sultry day in a community garden in Houston enables gardeners to trade stories, and volunteer gardeners often bring a share of their bounty of vegetables and herbs to elderly housebound residents living nearby.<br><br>Friends of the Rouge Watershed, a civic ecology organization outside the city of Toronto, sponsors interfaith Muslim-Jewish tree planting days, and advocates for formal recognition of the watershed they are restoring as the world's largest urban park. |

In including civic engagement as a component of social capital, Putnam harkens back to the work of the French observer of nineteenth-century America Alexis de Tocqueville. Tocqueville portrayed voluntary associations as "schools" of public spirit, civic mindedness, and democracy that helped people recognize obligations to their larger community.[25] And because voluntary associations like farm women's sewing clubs, fraternal Elks Clubs, parent-teacher associations (PTAs), or even bowling with friends on a Saturday night, have declined in popularity as people retreat into their homes and spend hours in front of their TVs and computers, Putnam was concerned not only about the future of civic obligations, but also about the very foundations of our American democracy.

Yet other scholars of civic life are less concerned. They report that older forms of civic engagement are being replaced by contemporary forms, and that in fact, we absolutely need to replace traditional forms of engagement. Previously, social capital was often built through exclusionary gender-specific or race-segregated organizations like fraternal orders that excluded women, country clubs that barred Jews, or the secret and sinister Ku Klux Klan.[26] What's more, when the social bonds and trust that are part of social capital do lead to some sort of action, that action may not benefit the community. In impoverished neighborhoods surrounding the Guatemala City dump, or the high-rise housing projects in U.S. cities, violent gangs form close bonds and mutual trust that help to ensure their own survival, but these same bonds and trust may depend on gang members killing neighbors and even family members outside their gang. Hence the questions: What forms of social capital lead to positive outcomes for neighborhoods and the environment? And what, in addition to social capital, is needed to ensure that social trust and bonds are deployed for the public good?

Rather than the much-lamented "civic decline" of Putnam, civil society scholars Carmen Sirianni and Lewis Friedland talk about a "civic challenge." They submit that as society faces ever more complex problems, and ever more diverse players with increasingly divergent interests, we are creating much-needed "new organizational and institutional capacities for collaborative problem solving and public work." These capacities complement existing policy tools like regulatory enforcement and market incentives, which "prove inadequate unless coupled with new forms of civic trust, cooperation, deliberation, and learning."[27] Harvard's Robert

Sampson and his colleagues are more blunt in their critique of traditional forms of civic engagement. They ask: "Why should we care about the prevalence of membership in the Elks Club, PTA, or any other group, unless it translates into collective civic action? Is it important that fewer people trust 'generalized others' or attend regular meetings? Does this tell us about the capacity of collectivities to act?"[28] Sampson and his Harvard colleagues propose a third notion—collective efficacy[29]—that along with a sense of community and social capital helps illuminate the civic component of civic ecology practice.

## Collective Efficacy: Change from Inside the Community

In 1962, the New York Housing Authority kicked off the Annual Tenant Gardening Competition. By the mid-1970s, over fourteen thousand residents in low-income housing projects were cultivating small plots adjacent to their buildings and entering the competition.[30] Soon it became evident to Charles Lewis, a horticulturalist assigned to judge the competition, that the importance of the gardens lay not only in the flowers. Not surprisingly based on what we have already learned from Lyda Hanifan, Robert Putnam, and others, gardening also provided an opportunity for residents to get to know each other, and the gardens became sites for social activities, weddings, and graduation ceremonies. In observations that foretold findings from research conducted two decades later, Lewis recounts: "After a few years the program started to produce unexpected results. Public housing tenants who grew flowers and vegetables joined together in a warm kind of pride and neighborliness. They saw to it that the gardens and surrounding areas were kept clean and neat. They urged the groundsmen to mow lawns adjacent to gardens more frequently."[31]

The gardening program spread to other cities. In low-income neighborhoods in Philadelphia, residents planted window boxes. And at the ill-fated Robert Taylor Homes in Chicago, tenants whose spirit had not been broken by gangs and disrepair planted a garden, decorated the walls in their shabby lobby, and painted drab entrance pillars red and white to match the colors of their flower garden.[32]

But the efforts of the gardeners resulted in more than social get-togethers, tidy lawns, and brightly decorated lobbies. The social activities and cleanups laid the groundwork for something larger. Residents no longer threw debris out their windows. They organized window watches to sound the alarm and mobilize residents when someone threatened to pick their flowers. And they assigned troublemakers among the children to guard their treasured plantings from vandals. In short, what started as planting flowers grew into cleaning up the common areas and then protecting newly prized gardens from crime—all of which required cooperation and organizing. What was it that led the residents of public housing to act for the common good—to push back against other forms of cooperation and organization like gangs?

Jump to the 1990s. Working in low-income neighborhoods in Chicago not unlike those Charles Lewis had observed over twenty years earlier, Robert Sampson and his colleagues focused not just on criminal activity and violence but also on smaller acts they called incivilities. Incivility refers to rude behavior. But it has broader implications than simply one person being rude to another. It also signifies acts "not conducive to the common good."[33] In some neighborhoods, residents are daily affronted by multiple incivilities—shattered school windows left unrepaired, rat-infested dumpsters, and young men loitering on the street corner and harassing passersby. All of these affronts to everyday life communicate an attitude of not caring, helplessness, and even hopelessness. This can lead to further disorderly behavior and more serious crime.[34]

One solution to the problem of incivilities is guided by the "broken windows" theory.[35] It suggests that if a community and its police force can fix the broken windows, remove the graffiti, clean up the vacant lots, and arrest the unruly teenagers, neighborhoods will become safer. Starting in the 1970s, police were instructed to come out of the patrol cars that isolated them from the people they were trying to protect, and instead were assigned to walking beats. They got to know the members of the community as well as who were the outsiders and possible troublemakers. And they confronted people for small "quality of life" crimes, or simply for infringing on community norms—like harassing people at a bus stop, jumping the subway turnstile, or throwing a rock through a window. The foot patrols helped to clean up the streets and made many people feel safer. In some instances, the police actions reflecting the mayors' fixing broken windows policies also reduced crime.[36]

But broken windows were only part of the story of incivilities and violence. In the 1990s, Sampson and fellow social scientists noted that

Chicago neighborhoods with the same levels of poverty and similar ethnic composition had different levels of violence. They wanted to know why. They organized a "drive-by" study. SUVs equipped with cameras on both sides cruised through low-income neighborhoods in Chicago's South Side and videotaped all the activity and buildings along the way. They captured the incivilities, as well as something else that they hadn't expected. And they compared what they saw on the videos with neighborhood crime and other statistics.[37]

It turned out that neighborhoods with lower crime rates were not necessarily the ones where someone from the outside came in and cleaned up the streets. Instead they were the neighborhoods where residents were meeting in a local church or school and figuring out how to solve problems together. Sometimes this meant intervening when teenagers were left unsupervised on the street corner. And other times it meant planting a community garden. Echoing Hanifan nearly a century earlier, Harvard political scientist Felton Earls described the significance of initiative coming from within the neighborhood.

If you got a crew to clean up the mess, it would last for two weeks and go back to where it was. The point of intervention is not to clean up the neighborhood, but to work on its collective efficacy. If you organized a community meeting in a local church or school, it's a chance for people to meet and solve problems.
If one of the ideas that comes out of the meeting is for them to clean up the graffiti in the neighborhood, the benefit will be much longer lasting, and will probably impact the development of kids in that area. But it would be based on this community action—not on a work crew coming in from the outside.[38]

Converting blighted lots to community gardens, it turns out, was one way for people to work together and thus create more long-lasting benefits than hiring an outside crew to clean things up. The community gardeners in turn were able to "reap a harvest of not only kale and tomatoes, but safer neighborhoods and healthier children."[39] Perhaps this is not surprising given that community gardens have been shown to serve as training grounds for participation in civil society.[40]

The Chicago study identified the willingness of neighbors to intervene for the common good. Collective efficacy is the name the researchers gave to this uniting of social cohesion and trust with—of utmost importance— the readiness to take action when things are going wrong.[41] Further, the researchers drew a connection between personal and collective efficacy, both having to do with the belief that taking action will produce desired outcomes. (We take up personal or self-efficacy in civic ecology practices in chapter 6 on health and well-being.)

## Summing Up

Collective efficacy emphasizes the intentional actions of individuals for the good of the community. Sampson would ask: In what communities are people willing to tell a child on the street to stop harassing passersby? Ostrom's emphasis is slightly different—she would want to know: What are the factors—such as social connections and how trustworthy you perceive your neighbors to be—that influence people to voluntarily limit their use of a common resource, and to manage that resource for the good of the community? Civic ecology practices combine notions of people taking the initiative to clean up incivilities in their community, and people collectively managing land, water, and other commonly held resources for the public good.[42] In this way, civic ecology links a long tradition of civil society scholars dating back to Tocqueville and passing through Robert Putnam, Robert Sampson, and Elinor Ostrom, to the writings of conservationist Aldo Leopold, who connected citizenship, stewardship, and community. Leopold wrote about humans as citizens and stewards of a "land-community," a community whose boundaries "include soils, waters, plants, and animals, or collectively: the land."[43]

When villagers in the Guatemalan highlands migrated to Guatemala City, or southern blacks in the United States migrated to northern industrial cities, they left behind a community associated with a place. Longing for a lost sense of community and wanting to belong, the young men joined gangs.[44] In this way, they tried to reestablish a sense of community, of belonging, and influence.

But the ties and trust that are part of social capital develop through healthy interactions as well as through unhealthy interactions in gangs. The willingness to join a voluntary association—or a civic ecology practice—to promote the common good suggests a focus on the positive aspects of how social capital develops and is deployed.[45] Closely related is the willingness of people to intervene when a child is throwing trash out the window, or a teen is defacing a public building. This willingness to intervene suggests that a community has collective efficacy—the capacity to take control of its own future.[46] Civic ecology practices provide a sense of belonging, they build social capital and are further enabled by social capital, and they demonstrate a community's willingness to engage in volunteer activities and to act for the common good. In chapter 10, we take up the issue of how governments and nonprofit organizations can help foster sense of community, social capital, and collective efficacy, through their policies that support civic ecology practices.

# Steward Story: Kazem Nadjariun

Kazem Nadjariun founded the volunteer group Nature Cleaners in Iran.[1]
He is retired now, but worked a number of different jobs in Iran and
Germany. His last job was at a cosmetics factory, where he did his best
to reduce water and electricity use and to recycle waste. Here he tells his
story in his own words:

When I was a kid, my family used to rent a house in the countryside north of
Tehran. We spent three whole months of summer there every year. It was a beautiful village and I played with a local child and discovered the area. My family did
that until I was twelve. After that, I often traveled to rural areas close to forests
on weekends. In this way, I became familiar with nature and [those] memories
remained in my mind. When I was a teenager, I traveled to the nature with my
friends but at that time I was not concerned about nature protection in a system-
atic way.

I left Iran when the Shah was in power and I moved to Germany. There, I
became familiar with environmental problems. Then I took a trip back to Iran
and I visited the village where we used to spend our summers when I was a kid.
I noticed how the village had been changed in a negative way. Gradually, I began
to face environmental problems in society in a more conscious way and my sympathy toward nature grew.

I came back to Iran fifteen years ago with my German wife and three children.
In Germany, I saw the positive effects of NGOs in protecting nature. After coming
back to Iran, I saw the awful situation of ecology and environment in my county
and it made me so sad. From the very beginning, my family and I tried to clean the
area whenever we went to the countryside on our trips. My German wife lives in
Iran and has a deep love for our country and its nature. She received a certificate
and a prize from the Sport Racing Federation of the Islamic Republic of Iran, in
recognition of her individual efforts to clean up car race tracks twelve years ago,
and she was featured in a television report.

My concern about our beautiful nature in Iran was raised even further when I
returned to the northern part of the country, and I saw lots of trash near a spec-
tacular lake called "Chorat." After cleaning the area with our friends, I was still
deeply concerned. Two days later, I decided to establish the group "Roftegarane

Tabiaat" (Nature Cleaners). My goal was to do the work in a group so that who-ever is concerned about cleaning nature can join in, and we can have an impact on other people in society. Also, so that we can keep our heads up when it comes to environmental protection issues in international society, and even can be a role model for other countries.

During the past one and a half years, Nature Cleaners has removed more than 100 tons of trash by holding 570 cleaning events all over the country. In addition, we post environmental awareness and educational materials on our online pages in order to make people aware about the environmental problems. This has a positive effect and causes more NGOs to care about our nature.

My work with Nature Cleaners has made me more aware of ecological prob-lems and the solutions for them. And it has made me even more determined to save nature.

# 4

## Oyster Spat and Live Oaks—Memories

*Civic ecology stewards draw on social-ecological memories to recreate places and communities.*

Roy's a south-shore boy. . . . His father was one of the biggest oyster-bedders in Prince's Bay—lost everything when they condemned the beds, and took a bookkeeping job in Fulton Market and dies of a stroke in less than a year; died on the Staten Island ferry, on the way to work.[1]

He's got an oyster tattooed on the muscle of his right arm. That is, an oyster shell. On his left arm, he's got one of those tombstone tattoos—a tombstone with his initials on it and under his initials the date of his birth and under that a big question mark.[2]

A tattoo embodies in Roy Poole's flesh memories of his father and the oysters' heyday. In the early 1950s, the men who had cultivated oysters—the "bedders"—still carried memories of a time when oysters abounded in the New York estuary. They also remembered the early 1900s, when sewage flowing into the harbor made the oysters unsafe to eat and closed down the oyster industry. And they remembered how making sure the oysters were "sleeping" before harvesting ensured that they would avoid diseases carried in the estuary's contaminated waters.[3]

When it became evident that oysters had become a hazard to human health, the bedders were allowed to transport the surviving shellfish to clearer waters off the coast of Long Island. But they didn't get them all: "A few were missed and left behind on every bed. Some of these propagated, and now their descendants are sprinkled over shoaly areas in all the bays below the Narrows."[4]

Just as the populations of oysters plummeted, so too did New Yorkers' memories of abundant oysters fade over time. But Mark Kurlansky, in his historical account *The Big Oyster: History on the Half Shell*,[5] helps us remember. The title of Kurlansky's book refers to the city's bygone nickname—the Big Oyster. At one time, oysters were so plentiful in the lower Hudson River and nearby waters that six cents bought all you could eat. Kurlansky described the bounty that characterized New York's estuary during a radio interview.

It's a kind of a lost history, but New York was an oyster center. It was always associated with oysters. In past centuries, if you were to say, I'm going to New York, the most likely response would be, enjoy the oysters. It was known for oysters. Oysters were sold in stands on every street corner. You could go to these huge open markets downtown any time of night and get oyster stew or raw oysters.

All five boroughs produced oysters. There were enormous oyster beds. New Yorkers never think about it this way, but New York City is the estuary of the Hudson River and it's this vast estuary of tidal basins and marshes and grassy wetlands and places that are perfect for growing oysters. And they did by the millions and ate an enormous amount of them and also sold them all over the world.[6]

Today, although the bounty and the bedders are gone, the estuary still harbors a small, scattered "biological memory" of these curious bivalves. The remnants of a once seemingly inexhaustible supply of oysters are sparsely "sprinkled over shoaly areas in all the bays."

### Biological Memories, Social Memories

Biological memories consist of the biological or genetic material needed to recolonize an ecosystem. They are the remnant populations of plants and animals that civic ecology stewards use to restore broken places. If oysters had become extinct—their biological memory forgotten—restoring them, alongside the important functions they contribute like food, filtering contaminants from water, and protecting fragile coastlines—would not have been possible.[7]

Social memories are shared among a group of people and help shape our behaviors—including our actions as members of a community.[8] Residents of New York still retain social memories of healthier city waterways. The bedders also carry social memories of how to cultivate oysters. These memories have been carried down several generations through the stories told by bedders to their children and grandchildren, and documented in books by Mark Kurlansky and Joseph Mitchell.[9] They help explain current day oyster gardeners' stewardship actions.

Today, New York's oyster gardeners are using biological memories— the remnant oysters—along with social memories of oyster cultivation.

Together with scientists who are researching the most effective means of raising shellfish, the volunteers cultivate the young oysters, called spat, and place them in artificial cages or reefs in the Bronx and Hudson Rivers. While slowly reestablishing the shellfish populations, these efforts also provide the "spats" of hope that someday the city once again will harbor bountiful and healthy oysters, along with the fish, birds, and other wildlife that thrived alongside the abundant shellfish. When asked about eating the oysters, the young people who are helping to place artificial reefs in the Bronx River respond with a sense of hope peppered with reality: "No, we cannot eat them today. But they are filtering the pollution out of the water and someday we will be able to eat the oysters."

## Memories across Time and Place

Like New York, the city of Stockholm harbors a vast estuary, its waters sparkling on a rare sunny day. And like New Yorkers, residents of Stockholm are applying biological and social memories in civic ecology practices. Only this time, the practice is allotment gardening instead of oyster gardening.

Similar to community gardens in the United States and Canada, allotment gardens are places where people cultivate plants in cities. Some allotment gardens in Stockholm are over a hundred years old. Gardeners grow long-lived fruit trees and raspberries that form hedges and attract pollinating bees. They grow strawberries transplanted from their grandfathers' gardens, and flowers the seeds of which have been passed down for generations. They recount stories about growing techniques told by their grandparents. And they also recreate elements of the places they remember—like small garden chalets that bring to mind traditional farmhouses. Each allotment is separated from the neighboring plot by a low fence or hedge that enables gardeners to peer over and see what is growing "next door." In this way, gardeners use not only memories of practices passed down from their own relatives, but also those from the forebears of their fellow gardeners.[10]

But memories not only are transmitted over time, they also are carried across space. The Stockholm allotment gardeners' memories of gardening practices passed down over generations are being employed in places distant from the farms where they originated. In New York, African

American gardeners cultivate cotton, a plant they remember from the farms they grew up on in the rural south. Sometimes social memories of planting practices are carried not just when people move from farms to cities, but when they flee across oceans. Hmong refugees who garden in Sacramento retain memories of farming practices from Laos. They have carried these memories first to refugee camps in Thailand, and then across the Pacific Ocean to California, and are using them to inform the way they grow gardens in a new land. Perhaps they also carry memories of how the act of cultivating life aids in recovery from witnessing death and from experiencing unspeakable hardship.

Similarly, biological memories are transported from one place to another. Hmong refugees from Laos plant a colorful panoply of gourds, squashes, eggplants, and beans in Sacramento using seeds carried from their homeland. Although oysters were present in the New York estuary in the late nineteenth century, to meet the demand for larger oysters, oystermen transported the spat of bigger varieties from the Chesapeake Bay.[11] Thus, similar to social memories of cultivation practices, the biological memories used in a civic ecology practice may come from outside the system being cultivated or restored. Sometimes, communities purposely set aside reserves that serve as a source of biological memories—animals and seeds that can move into new locations—such as fish in marine reserves, or trees that are retained in sacred forests in India so as not to not lose all possibility of planting new forests.[12]

## Social-Ecological Memories

When memories of gardening, tree planting, and oyster cultivation also include seeds, seedlings, cuttings, and oyster spat, they are referred to as social-ecological memories. Such memories constitute the library and the tools that enable recent immigrants, as well as gardeners several generations removed from farming or oystering, to fill vacant spaces in cities with plants and wildlife, and with community and cultural activities. In Stockholm, Sacramento, and New York, social-ecological memories—memories of cultivation practices and of seeds, cultivars, and spat—shape the allotment, community, and oyster gardening practices.[13]

Where communities are directly dependent on hunting, fishing, or growing food, social-ecological memories can be critical to survival and preserving a way of life. In the early part of the twentieth century, Cree hunters, armed with newly acquired repeating rifles, slaughtered hundreds of caribou. The caribou herd on which they had depended for food and cultural traditions disappeared from Cree hunting land. Seventy years later, caribou from neighboring areas—areas not unlike marine reserves set aside to maintain fish populations—reentered Cree territory. Although initially, younger hunters who did not hold memories of the catastrophic events two generations earlier again slaughtered the caribou, the elders' retelling of their tragic stories eventually spurred the younger Cree hunters to adopt more sustainable practices.[14] The survival of the Cree's subsistence way of life was made possible by the community's social-ecological memories—the elders' memories of traditional hunting practices along with the remnant caribou herd.

As evidenced by the oyster bedders, allotment gardeners, and Cree hunters, social-ecological memories accumulate over multiple generations of resource users. As the places they inhabit experience droughts, floods, temperature fluctuations, contamination, insect outbreaks, and other sorts of change, bedders, gardeners, and hunters are continually trying out new practices, and observing which practices lead to fast-growing oysters, healthy vegetables and fruits, and abundant wildlife. Through such observing and adapting over generations, fishermen, farmers, and hunters accumulate social-ecological memories that carry a wealth of knowledge and experience. When passed down over many generations living in the same location, memories embody a rich knowledge of the local ecosystem and how it has changed, and of the outcomes of different management practices. These memories in turn enable a community to respond when faced with new changes or stress. In this way, social-ecological memories are not simply "a stagnant pool of knowledge" from bygone days, but a source of information and even innovation that allows for "the recasting of core ideas from a deeper past so they can be used to respond to the new circumstances."[15]

## Memories, Defiance, and Recovery

The experience of the Cree caribou hunters demonstrates how social-ecological memories played a role in a community's recovery from a local catastrophe. In cities, social-ecological memories also play a role in recovery after catastrophic disasters, as well as from ongoing smaller disruptions.[16] Sometimes these memories include iconic species, such as Louisiana's live oak trees with their majestic spreading crowns and sinuous branches laden with Spanish moss. After Hurricane Katrina, social-ecological memories of New Orleans' iconic live oaks—and people planting new seedlings and caring for oaks that survived—became a symbol of hope for renewal of the city and its residents. This was especially true in the neighborhood called Tremé, where elderly residents retained social memories of how live oaks had shaped their neighborhood well before the disaster.

Claiborne Avenue runs through the Tremé neighborhood of New Orleans. Historically, Claiborne Avenue was lined with ancient and stately live oak trees. The oaks provided a public green space, which the area's mostly African-American residents used as a community gathering site. In the late 1960s, the highly contested construction of an elevated highway above the oldest section of Claiborne Avenue pitted New Orleans' wealthier neighborhoods against Tremé. Tremé lost. After construction, poorly lit cement parking lots under the freeway replaced grassy gathering grounds, and concrete supports for the highway took the place of oak trees. Construction of the overpass is seen as an important factor leading to the overall decline of the Tremé neighborhood in the 1960s and 1970s.[17] In 2002, as part of the *Restore the Oaks* art installation, painters recreated symbolic live oaks on the freeway columns. In this way, the community was able to memorialize the symbolic trees, as well as the thriving social life that once characterized Claiborne Avenue.

After Hurricane Katrina in 2005, memories of the Claiborne Avenue highway construction, and of the trees and neighborhood cohesiveness that were subsequently lost, reemerged. Surviving residents of Tremé began planting trees with a renewed vigor and sense of urgency—an urgent biophilia. A community elder recalled: "We remember, just about five short blocks from here, we have Claiborne Avenue, which was a beautiful corridor of oak trees that, it's unfortunate, but the government came through with the interstate, and they knocked all the trees down . . . it destroyed the neighborhood; by destroying two hundred or three hundred year old trees, they destroyed the neighborhood. We need to do the opposite of that."[18]

In Tremé and other neighborhoods decimated by Katrina, planting and caring for trees became an act of defiance. The residents refused to let

*Restore the Oaks* art installation under the Interstate 10 highway in Tremé neighborhood of New Orleans. Photo credit: Jean Fahr, Parkway Partners

their memories of an earlier Claiborne Avenue fade, and they used those memories to defy the years of neglect, the more recent catastrophic destruction, and the media's dismissive claims that New Orleans lacked the capacity to rebound. Seeing reports of other defiant tree planting projects on the news, and remembering the live oaks of bygone times, residents in Tremé realized that they did not have to let the new disaster destroy their neighborhood, as the freeway had done a generation earlier.[19]

In New Orleans, social memories encompass not just how to care for trees, but also how trees help us form emotional attachments to the places where we, and our forebears, have lived.[20] Similarly, by the very act of planting yard-long beans, Thai eggplants, and other biological memories from their homeland, Hmong refugee gardeners in Sacramento are doing something more than simply using social-ecological memories to cultivate a garden. They also are connecting to the places and people of their earlier homeland. In their own way, they are creating living memorials to their past—and acts of defiance that symbolize their ability to survive and move beyond the horrors of war, holocaust, and diaspora.

### Recreating Knowledge, Expressing Memories

A tattoo on the arm of an oyster bedder's son embodies memories of an estuary where oysters once flourished alongside the bedders who culti-vated them. As the memories of the shellfish and their cultivation faded, much of the knowledge about how to raise oysters, and about their role in the culture and traditions that once symbolized New York, was lost. Where social and biological memories are depleted, for example when a generation of elders holding a particular type of knowledge has died, or a species has gone extinct, social-ecological systems have fewer resources to draw from as they deal with change and uncertainty.

Knowledge of oyster cultivation is being recreated as volunteers, high school students, scientists, and resource managers are coming together to restore the oyster's habitat. They are experimenting with artificial reefs of different sizes and shape. They are trying out spat from different genetic sources. And they are accumulating new knowledge each time a shellfish steward tries to design a slightly better reef or uses a different source of spat. Through time, this knowledge will become incorporated into the social-ecological memories that guide future oyster cultivation practices.

In this way, not only are memories used to inform civic ecology prac-tices, but the practices themselves create knowledge that will become new social-ecological memories.[21] Thus, similar to the situation with social capital we described in the previous chapter, memories are both deployed to shape a civic ecology practice and developed through these practices. Like social capital, social-ecological memories in civic ecology practices demonstrate circularity, or feedbacks. As we see throughout our explora-tion of civic ecology practices in the chapters of this book, such relation-ships are common in social-ecological systems—perhaps more prevalent than linear, one-way relationships.

As they create new memories of cultivation practices, the oyster gar-deners in New York, tree planters in New Orleans, allotment gardeners in Stockholm, and community gardeners in Sacramento are also re-creating New York, New Orleans, Stockholm, and Sacramento as different places. New York is not just a city of skyscrapers, it is also a place where families can walk along blossoming waterfronts and paddle in rivers and wet-lands. New Orleans was not wiped off the map, but rather it is a place where people are reconnecting with each other and with the trees that symbolize their history and their city. Stockholm is a network of green spaces, including allotment gardens, cemeteries, and municipal parks, which together provide habitat for pollinating insects and for birds. And Sacramento is a place that welcomes refugees along with the cultures they bring to their new homeland. Let's take a break to explore some of these places through a civic ecology photo gallery.

# Gallery of Places

Civic ecology stewards restore and transform broken places. Often they cultivate symbols of places they have left behind or create new defiant symbols of place. Here we present a gallery of civic ecology places. Can you tell where each is from? (A key is included along with the photo credits at the close of the gallery.)

Photo 1

Photo 2

Photo 3

Photo 4

Photo 5

Photo 6

Photo 7

Photo 8

Photo 9

Photo 10

## Gallery of Places Credits and Key

**Photo 1**
Volunteer stewards from Surfrider Foundation rebuild sand dunes at Rockaway Beach, New York, after Hurricane Sandy. Photo credit: Bryce DuBois, Civic Ecology Lab, Cornell University

**Photo 2**
Students from the community organization Youth Ministries for Peace and Justice monitor water quality in the Bronx River. In the background is a former concrete plant, which, through the efforts of the Youth Ministries for Peace and Justice and myriad other government and civil society organizations, has been transformed into Concrete Plant Park. Photo credit: Alex Russ, Civic Ecology Lab, Cornell University

**Photo 3**
Although we often think of civic ecology practices as taking place in cities, they emerge in other stressed communities and environments. This one is from the Eastern Pennsylvania Coalition for Abandoned Mine Reclamation (EPCAMR), which restores and monitors stream runoff from abandoned coal mines. It shows an artesian flow of abandoned mine drainage discharging from a borehole that was intentionally drilled in 1974, two years after the historic 1972 Agnes Flood. The borehole was intended to relieve the pressure and flooding of basements and other areas caused by Agnes and mine water filling underground mine pools. Shown is Robert Hughes, EPCAMR executive director, holding an electro-shocking probe. Hughes was conducting a fish survey upstream and downstream of the

boreholes with members of Trout Unlimited. The data were used to help plan for future reclamation efforts. Photo credit: EPCAMR

**Photo 4**
What is now Soweto Mountain of Hope (SOMOHO) was a dangerous and deserted wasteland during apartheid. In recent years, township residents have transformed SOMOHO into vegetable and AIDS memorial gardens, a playground, theatre, sculpture display, and community center running education and health programs. During the 2002 United Nations World Summit in Johannesburg, UN Secretary-General Kofi Annan visited SOMOHO to showcase the garden as an example of sustainable development. Photo credit: Zute Lightfoot

**Photo 5**
Hmong refugee growing foot-long beans in a community garden in Sacramento, California. Photo credit: Mark Whitmore

**Photo 6**
Youth from Rocking the Boat check on the oysters they are cultivating in the Bronx River. The boat, a replica of the wooden rowboats that once plied the New York harbor, was constructed by youth in Rocking the Boat's woodworking program. Photo credit: Adam Green/Joaquin Cotten/www.rockingtheboat.org

**Photo 7**
In the Bronx Community and Cultural Garden, young people learn about culturally significant planting practices from Puerto Rican elders. Photo credit: Alex Russ, Civic Ecology Lab, Cornell University

**Photo 8**
Demonstrating hope for the future, volunteers in Detroit plant trees in front of a boarded-up house. Greening of Detroit conducts many such tree plantings on vacant properties around the city. Photo credit: Greening of Detroit

**Photo 9**
Student members of Friends of the Gorge repair a trail along a gorge on the Cornell University campus. Friends of the Gorge is Marianne Krasny's personal civic ecology practice. She started it after a series of student suicides and accidental deaths in campus gorges, and because of her love for the gorges, developed over many years. Photo credit: Mark Whitmore, Cornell University

**Photo 10**
Canoga Creek Farms, of Canoga, New York, encompasses a wetlands restoration project. Volunteers plant wetland grasses using hydro-seeding. Canoga Creek Farm's conservation program is Keith Tidball's personal civic ecology practice. Photo credit: Keith G. Tidball, Civic Ecology Lab, Cornell University

# 5

## Ecosystem Services: What's Nature—Including Humans—Got to Offer?

*Civic ecology practices produce ecosystem services.*

Biodiversity, understood broadly as the living component of ecosystems, is at the core of human well-being, because it affects, and often underpins, the provision of ecosystem services.[1]

### Biodiversity and Ecosystem Services: A History and a Controversy

In the early 1980s, conservationists grew increasingly alarmed. Despite any evolutionary-based love of life as had been proposed by E. O. Wilson in his notion of biophilia,[2] humans were forcing other species into extinction at an unprecedented rate. Moreover, nobody seemed to care. People watched on television as fires, chainsaws, and bulldozers transformed biodiversity-rich Amazon forests into dry cattle ranches depauperate of other species. In the United States, it appeared that the spotted owl and the last vestiges of old-growth Pacific Northwest forests would be sacrificed to ever greater demands for wood and paper products. And yet these and other disquieting losses of biodiversity, the outcome of millions of years of evolution, were not spurring people to action.[3]

It was at this point that conservation scientists Paul and Anne Ehrlich of Stanford University hit upon another idea—a way of communicating about the utilitarian value of Earth's biodiversity that they hoped would force people to take note. They proposed the term *ecosystem services* as a metaphor to help people grasp the gravity of what was being lost.[4] Were people to understand that biodiversity provided important services to humans, services that would cost billions if not trillions of dollars to replace if we could even find substitutes, perhaps they would be quicker to support conservation. The notion of ecosystem services—the services that humans derive from the rest of nature—would be used to educate people about the impacts of biodiversity loss on themselves as individuals and on humanity as a whole. And it would transform human behavior.[5]

In 1997, a group of top thinkers from the new discipline "ecological economics" published a highly influential—and controversial—paper in which they calculated a dollar figure for the total global worth of the Earth's ecosystem services. The authors summarized their conclusions:

The services of ecological systems and the natural capital stocks that produce them are critical to the functioning of the Earth's life-support system. They contribute to human welfare, both directly and indirectly, and therefore represent part of the total economic value of the planet. We have estimated the current economic value of 17 ecosystem services for 16 biomes, based on published studies and a few original calculations. For the entire biosphere, the value (most of which is outside the market) is estimated to be in the range of US\$16–54 trillion ($10^{12}$) per year, with an average of US\$33 trillion per year. Because of the nature of the uncertainties, this must be considered a minimum estimate.[6]

This paper fueled a vigorous debate about ecosystem services. What had been viewed as a tool for communicating about the human consequences of loss of biodiversity now acquired a second—economic—meaning. Economists and ecologists set out to determine how to more accurately calculate the economic value of services provided by the Earth's organisms—its biological diversity—to humans.[7]

Alarmed by this new focus on economics, other scholars jumped into the fray. They contested the notion that meaningful dollar values could be placed on services offered by nature, and claimed that monetizing nature was in fact unethical (and antithetical to values about life that humans should hold dear).[8] An economic approach, they argued, would foster damaging motivations for conservation. It would promulgate a counterproductive worldview—that nature exists solely to produce services for human consumption.[9] Others worried that monetization would ignore noneconomic cultural values, especially in indigenous societies whose traditions regarding nature differed radically from those of westerners.[10] More fundamentally, a feeling of unease with economic valuation gnawed at those who know in their hearts that nature is about meaningful connections, about connecting with our evolutionary—and ecological—selves.[11]

A more recent set of concerns parallels a shift in thinking from ecosystems to social-ecological systems. Scholars adhering to a social-ecological systems view felt that economic measures would be incapable of accounting for the dynamic linkages between human social and ecological processes in the production of ecosystem services. They claimed that ecosystem services are produced by complex social-ecological systems and not by ecosystems alone.[12]

Despite these concerns, the focus on monetary valuation has played a role in mainstreaming ecosystem services and garnering political support for conservation. In 2005, over a hundred scientists joined together to produce the *Millennium Ecosystem Assessment,* outlining the importance of four different categories of ecosystem services to human health and community well-being, and detailing the speed at which services were being lost (see box below).[13] And in 2013, an international group of

Four Types of Ecosystem Services

---

The *Millennium Ecosystem Assessment*[14] describes four categories of ecosystem services:

*Provisioning services supply the basics that humans need to survive.* They include the food we eat, the water we drink, the fiber we make into our homes and clothes, fuel to keep us warm and cook our food, the natural medicines and pharmaceutical drugs we derive from nature, as well as the genetic resources that could be the source of future food, fiber, and medicines.

*Regulating services regulate ecosystem processes.* They mitigate air, water, and solid waste pollution, climate extremes, flooding, erosion, diseases, pests, and natural hazards. Pollination is also considered a regulating service because insects and other pollinators regulate the diversity of plant species and the amount of food we are able to grow.

*Supporting services differ from the first two categories because their impacts are longer-term.* They are things like soil formation, photosynthesis, and nutrient and water cycling, which support the other services rather than provide direct benefits to people.

*Cultural services are the nonmaterial benefits that people derive from nature.* This category includes opportunities for cultural, spiritual, inspirational, aesthetic, educational, and recreational experiences provided by diverse ecosystems.

All ecosystem services have direct or indirect benefits for human health and community well-being. For example, climate and flood regulation proffer protection from disaster. Recreational, spiritual, and stewardship experiences in nature foster psychological, emotional, and physical health. Civic ecology practices are a source of multiple ecosystem services.

scholars produced the first global assessment of urban biodiversity and ecosystem services.[15]

In putting forward the notion of ecosystem services thirty years ago, Ehrlich and Mooney were concerned about people who had no clue as to the value of natural ecosystems. Civic ecology stewards, in contrast, implicitly exhibit such an understanding—through their stewardship actions, they demonstrate a commitment to conservation. They are restoring land, water, and biodiversity similar to how conservationist Aldo Leopold transformed an abandoned farmstead into productive prairie and forested woodlots.

Despite the fact that civic ecology stewards demonstrate through their actions that they intuitively value what ecosystems provide, they may still benefit from familiarity with the notion of ecosystem services—it can help them to understand, monitor, and communicate the outcomes of their stewardship actions. The notion of ecosystem services provides a framework for civic ecology stewards to answer the questions: What am I providing for my community and environment through my stewardship actions? Am I providing what I set out to provide? And how might I do better?

As a start to answering these questions, we explore how a Japanese traditional landscape incorporating diverse land uses produced a range of ecosystem services. And we discover efforts by the Japanese to restore this landscape, its biodiversity, and its ecosystem services. Following our sojourn to Japan, we turn to how ecosystem services are being produced and monitored in cities, including in civic ecology settings.

### Satoyama and Dragonfly Ponds

In Japanese samurai culture, the dragonfly is called the "victory insect." It is revered as an animal that flies only forward and does not retreat. For thousands of years, Japanese farmers also held an uneasy reverence for the samurai class. And they watched the fearsome dragonflies darting back and forth over their rice fields and swooping in to nab mosquitoes in mid-air. Today, the Japanese pride themselves on being able to recognize the 180 species of dragonflies inhabiting the archipelago,[16] not unlike bird watchers in the United States who like to show off their prowess in identifying numerous species of birds.

The Japanese farmers managed the landscape so that it produced food and shelter, as well as provided the wetland habitat critical to dragonflies. They adjusted the water levels in streams, ponds, and reservoirs to create wet rice paddy fields inhabited by fish, frogs, and birds, in addition to dragonflies. They maintained forests at different stages of succession by periodically cutting trees, which in turn provided sunlight for wildflowers. They also collected wood for buildings and fuel, bamboo shoots for food, and leaves to compost and apply as fertilizer to the rice paddy fields. In the villages, people cut grass for thatch, bedding, fertilizer, and hay for their animals. Taken as a whole, this Japanese mosaic of different land uses—rice paddies, grasslands, woods, and villages—is called satoyama. Satoyama landscapes provided the ecosystem services that the farmers and villagers depended on.[17]

Starting in the 1960s, the introduction of pesticides and modern methods for growing rice that no longer depended on wetlands changed the

Restored Niiharu Satoyama site in Northern Yokohama, Japan. Sign refers to the name of the community group that restored the site. Photo credit: Keith G. Tidball, Civic Ecology Lab, Cornell University

satoyama landscapes. The mosaics of different patches of land uses disappeared. So did the children of farmers who increasingly abandoned the rural way of life for opportunities in the city. Farmers remaining were able to produce more food, but at a cost to their families, and to the dragonflies, birds, and wildflowers that had prospered in the satoyama landscapes.[18]

Toward the end of the twentieth century, spurred by their profound love and the cultural significance of dragonflies and other species, and concerned that a younger generation would lose important symbols of Japanese culture, people in Japan began to restore satoyama landscapes. One such project started in the Sayama Hills on the outskirts of Tokyo. Several thousand volunteers worked with farmers to rebuild paddy fields and water reservoirs and to grow rice using traditional practices. They also planted tree seedlings, mowed grasslands, thinned forests, and cut back forest undergrowth. Together their activities restored the ancient satoyama system. Today, children visit the sites to learn about wildlife, and about Japanese agricultural and cultural traditions.[19]

Not only in rural areas, but also within their densely built-up cities, the Japanese have found a way to bring back the beloved dragonfly. In Hommuku Citizens Park in Yokohama, a concrete-lined pond was home to ornamental fish but devoid of plants and frequented by only three dragonfly species. In 1986, citizens groups, scientists, and city government partnered to construct a winding stream with pools in both shady and sunny spots, bordered by newly constructed earthen banks to encourage native aquatic plants. The re-created habitat was soon the home to twenty-seven dragonfly species. The reintroduced dragonflies, in turn, attracted school children and dragonfly aficionados, who came to learn about nature. Visitors also helped steward the pond's biodiversity. They removed unwanted plants, dredged sediment, and captured the crayfish and introduced bluegill sunfish that prey on the dragonfly larvae. Within fifteen years, what started as one small pond restoration had sparked a movement—hundreds of dragonfly ponds had been created across Japan, many serving as sites for the public to learn about and help steward nature.[20]

Japan's satoyama landscapes and restored ponds in cities demonstrate how biological diversity and ecosystem services are connected. They are home to a diversity of plants and animals. Satoyama systems produce rice, fish, and bamboo shoots for food, wood for building, and fuel for heating and cooking—provisioning services (see box above for different categories of ecosystem services). And composting leaves is a supporting service, recycling organic material to create soil. Modern agriculture focuses more narrowly on producing rice—a single ecosystem service. Similarly, in cities, the concrete-lined ponds offered little in the way of biodiversity, whereas the dragonflies in the more species-rich restored ponds eat insects, thus providing a regulating ecosystem service. They also provide rich educational, recreational, and stewardship opportunities—that is, cultural ecosystem services—for the children and adults of Japanese cities.

### Do Cities Provide Ecosystem Services?

Cities are generally portrayed as rampant consumers of ecosystem services, and as responsible for paving over the forests, farm fields, wetlands, and even rivers that provide those services. Cities consume water, food, fiber, and energy. Instead of helping to regulate floods or nurture pollinating insects, their ubiquitous impermeable surfaces, including roads, parking lots, and roofs, worsen runoff and fail to provide habitat for bees and other pollinators. They have relatively few means of recycling waste, and their lawn chemicals, oil leaking from cars, and industrial contaminants flow directly into rivers and oceans. And as for recreation and spiritual renewal, city dwellers often travel to distant parks and protected areas seeking cultural services seemingly unavailable in surrounding urban environments.

Yet cities—and in particular their biodiversity and civic ecology practices—also can be a source of ecosystem services. In the South Bronx neighborhood of New York, youth contribute to a variety of ecosystem services. They construct artificial wetlands along the Bronx River and rooftop gardens on nearby buildings to reduce runoff. They grow mussels and seaweed on wooden rafts strategically anchored near combined sewage overflow pipes, as part of an experiment to develop better means of filtering out excess nutrients.[21] They plant vegetables, compost wastes, create art objects from colored tiles, and pull invasive species in community gardens. And they take families on tours of the Bronx River in wooden rowboats during community rowing days. Through their activities, the young people help to provide fresh food, pollination, water filtration, stormwater retention, soil formation, nutrient recycling, and recreational and educational opportunities for their South Bronx neighborhood.[22]

In Stockholm, Sweden, researchers compared ecosystem services produced in allotment gardens, cemeteries, and city parks. For the ecosystem service pollination, the allotment gardens outperformed the other green spaces. The allotment gardeners planted more flowers and actively protected bumblebee nests—hence their gardens' greater abundance of pollinators.[23] They also composted, fed birds during the winter, set out nest boxes and birdbaths, and protected species that preyed on insect pests. In so doing, they provided ecosystem services including nutrient recycling, natural pest control, and recreational opportunities such as bird watching and gardening. Studies of community gardens in the United States have similarly documented their contributions to pollination,[24] as well as to food production[25] and education.[26]

Yet despite studies showing that not only formal green spaces like city parks and cemeteries, but also civic ecology practices provide ecosystem services, municipal governments often have overlooked the contributions of these community stewardship initiatives to ecosystem services. Perhaps this is because community gardens and other civic ecology practices at times have had adversarial relationships with government. In some instances, the community gardeners were seen as "squatters" on land for which they didn't hold title. And in New Jersey, citizen-driven efforts to restore oysters in polluted waters were banned because of concerns that contaminated oysters might find their way into the commercial oyster supply.[27] In other cases, civic ecology practices and sites went unnoticed. They were small and scattered throughout a city—people walked right by and didn't look inside.

But nonprofit groups, mayors, and even popular entertainers are starting to take note of the contributions civic ecology practices make to quality of life and the environment. In New York, Bette Midler joined forces with Mayor Bloomberg to launch MillionTreesNYC—a partnership between the city and community groups to plant a million trees that will cool and clean the air and provide recreational opportunities. And New York is just one of many city governments forming coalitions with civic ecology groups to provide fresh food, educational opportunities, shoreline protection, and other ecosystem services.[28]

Youth from the community organization Rocking the Boat work with ecological artist Lillian Ball to plant WATERWASH ABC, a 30,000-square-foot artificial wetland along the Bronx River. The completed wetland treats 10,000 square feet of parking lot runoff from the adjacent retail store ABC Carpet and Home. Photo credit: Adam Green/Joaquin Cotten/www.rockingtheboat.org; A Green, "South Bronx Nonprofit Rocking the Boat Teams with Ecological Artist Lillian Ball, Edesign Dynamics, Drexel University and ABC Carpet and Home on Project to Reduce Pollution in Bronx River," http://longislandsoundstudy.net/wp-content/uploads/2011/10/WATERWASH-ABC-Media-Advisory.pdf

## Measuring: Do We Produce What We Say?

By partnering with community groups and businesses, Mayor Bloomberg's MillionTreesNYC claims that it will become an important provider of ecosystem services:

> Through a mix of public and private plantings, MillionTreesNYC, an important initiative of PlaNYC, will expand New York City's urban forest by 20 percent. All New Yorkers will share in the many benefits that come from planting trees—more beautiful neighborhoods and parks; cleaner air and water; higher property values; energy savings; cooler summer streets, yards, and public open spaces; and a healthier, more environmentally sustainable City. MillionTreesNYC will get New Yorkers involved in the planting and caring of trees for the next decade.[29]

Yet even as governments are recognizing the value of engaging with civic ecology practices, there is a danger that their enthusiasm will wane in the absence of proof that claims are valid, such as those made by MillionTreesNYC. Perhaps it is not surprising that relatively few civic ecology stewards monitor their impacts on biodiversity or ecosystem services.[30] Stewards, many of them volunteers, express their commitment to—and love for—nature and their community through active stewardship, but may be less inclined to undertake the oft-times tedious work of measuring air temperature, water runoff, or hours spent in recreation. They also may lack the funding and scientific expertise to monitor their contributions. And even if a leader of a community garden or tree planting effort is committed to and has the resources to document impacts, she may have limited authority to compel volunteers to follow along.

Despite these challenges, some civic ecology stewards have devised means for measuring ecosystem services. Friends of the Los Angeles River records pounds of trash collected during volunteer cleanups, an indicator of aesthetic values or cultural services.[31] And friends of parks groups who remove invasive species in Seattle have partnered with the nonprofit EarthCorps to map vegetation in the parks,[32] which provides an indication of their contributions to native biodiversity and to aesthetic and recreational experiences. In New York, community gardeners conduct "crop counts" of the number of vegetable or fruit-bearing plants and "harvest counts" of the weight of their produce. Composting for Brooklyn weighs the pounds of food scraps composted (8,600 pounds in summer 2011![33]) as a measure of the regulating service nutrient cycling.

Feedback Farms,[34] like most community gardens, occupies a formerly vacant lot. But it stands out among community gardens in that it faces head-on what many gardens experience as their single greatest struggle—lack of secure land tenure—and turns this liability into an asset. Knowing that any one vacant property is likely to be developed during an economic upswing, Feedback Farms plants vegetables in movable containers. If the group is kicked off its site by a new building going up, they can transport the containers to the next vacant lot. Each container is also a small-scale, urban agriculture experiment; gardeners vary the means of irrigation and the type of soil in the containers, and record the water added, soil moisture, and the hours of volunteer labor spent in growing crops with different water and soil treatments. All of these variables are then related to the weight of each of five vegetable crops produced in the containers—a measure of the food-provisioning ecosystem service. The gardeners at Feedback Farms use the results of their experiments to adapt their food-growing practices on an ongoing basis.[35]

In a hopeful sign for the future, freely available software, open technologies, and partnerships with scientists are helping civic ecology groups overcome barriers to measuring ecosystem services. Two examples are:

• NJ Tree Foundation, which has planted over 150,000 trees in Newark and other New Jersey cities, records the trees' species and diameters. They input the data into i-Tree and the software spits out dollar values for ecosystem services, including carbon sequestration, stormwater management, air quality, and summer cooling.[36]

• The Gowanus Canal Conservancy, located along the famed Brooklyn industrial waterway that has been designated a superfund site, has installed bioswale gardens and teamed up with a Drexel University scientist to monitor changes in water quality and quantity and soil contaminants.[37] In partnership with the Public Lab's "Red Balloon Adventure," the Conservancy also launches helium balloons equipped with low-cost infrared cameras. As they sail over the canal and adjacent Brooklyn neighborhood, the cameras capture snapshots of permeable and impermeable surfaces, tree canopy cover, and water plumes emanating from construction sites—all of which could be used for monitoring the impacts of civic ecology practices. And in a demonstration of how citizen monitoring can be combined with recreation, the Conservancy teamed up with the Gowanus Dredgers Canoe Club to monitor water quality while boating.[38]

### Civic Ecology Practices, Green Infrastructure, and Ecosystem Services

By planting a bioswale garden or installing an artificial oyster reef, civic ecology stewards help build green infrastructure in cities. This green infrastructure in turn provides ecosystem services such as water filtration or protecting coastlines from erosion. In fact, researchers studying ecosystem services in cities have focused on green infrastructure, such as trees, bioswale gardens, and even high-technology algal systems that produce biofuels.[39] But what about the participatory processes—the civic ecology practices—by which such services are sometimes produced?

Civic ecology stewards help create green infrastructure (such as a bioswale or community garden or an urban forest), which in turn produces ecosystem services like food or cooling. What's more, the stewardship practice itself can be a means of recreation, education, and aesthetic enjoyment, all of which are forms of cultural ecosystem services. Planting gardens is a form of recreation—thus community gardeners both produce recreational services and benefit from them. Laying down oyster reefs and seeding them with spat is a form of learning—thus oyster gardeners both produce educational services and benefit from them. And village groves along with their stone sculptures hold cultural significance—through restoring them, Korean elders both produce and benefit from aesthetic and spiritual services.[40]

In short, stewardship behaviors among humans produce ecosystem services both indirectly through creating green infrastructure and directly through the actual stewardship behaviors themselves. The importance of humans in creating the *green infrastructure* that provides ecosystem services, and of the participatory, self-organized civic ecology *practices that themselves contribute to ecosystem services*, suggests an expansion on the quote from the beginning of this chapter: "Biodiversity, understood broadly as the living component of ecosystems, is at the core of human well-being, because it affects, and often underpins, the provision of ecosystem services."[41] Whereas we all too often are faced with the destruction humans have wrought on biodiversity, it may be time to incorporate into our thinking how humans, like other parts of nature, also play a role in restoring biodiversity, green infrastructure, and attendant services.[42] *Biodiversity, and the efforts by groups of people to restore lost biodiversity, are at the core of human well-being, because biodiversity, green infrastructure, and stewardship underpin the provision of ecosystem services.*[43]

We started this chapter by stating that the notion of ecosystem services was intended to communicate how biodiversity contributes services that are essential to human health and well-being. We have seen that civic ecology practices, in addition to biodiversity, help provide ecosystem services. We turn next to the question: What are some of the health and well-being outcomes of being in, and of stewarding, nature?

# Steward Story: Helga Garduhn and Marian Przybilla

For years, Marian Przybilla and Helga Garduhn both taught school and advocated for environmental protection in Germany.[1] They brought nature to the students in the classroom and their students to nature outside. Garduhn founded a student ecology group in the basement of her home, and together with the children created a self-guided nature trail in the Breisetal Forest. Przybilla taught members of a youth group how to use an axe and a saw and how the forest served as a habitat for different kinds of life. Yet despite sharing common interests and living only a few kilometers apart, the lives of Garduhn and Przybilla did not intersect. Garduhn worked on the east side of the foreboding towers guarding the Berlin Wall, and Przybilla worked on the west side.

Then, at midnight on November 9, 1989, the "greatest street party in the history of the world" erupted, as East and West Berliners flocked to the Wall chanting "Open the gate!" People who became known as "wall woodpeckers" urgently grabbed hammers to bash the wall. And they clutched onto picks to chip away small, but symbolic, pieces of the massive barrier. Cranes and bulldozers joined in the demolition.[2]

Garduhn and Przybilla also felt a sense of urgency—an urgency to unite with their fellow Germans on both sides of the wall. They especially wanted to meet compatriots who shared their passion for the environment. Less than a year after the fall of the Berlin Wall, the two teachers arranged a meeting of their respective youth groups. From the moment they met, they knew they would understand each other. Not only did they share the same birthday, August 9, and professional lives as biology teachers, but also their intense desire to work for the environment. Together they would replant the former border strip and transform the Wall and its towers to something green and beautiful. Thus began a life-long friendship that was to change the face of the Berlin Wall.

Their initial idea was meant as a joke. Ms. Garduhn had been kicked out of her house by the new authorities, and thus no longer had a basement for her student ecology group to meet in. Together with Przybilla she submitted a request to the East German police: Can we use one of the old, now useless guard towers as a meeting place for our reunified student environmental groups? Unexpectedly, the East German police responded positively, and presented Gardhun with papers stating they had transferred the tower for use by her ecology group, but with the officious and out-of-place caveat that "Warranty service, spare parts delivery and repairs are not guaranteed." Although the upheaval in the aftermath of the fall of the Iron Curtain meant that the transfer did not go as smoothly as originally promised, by 1994, their odyssey through the bureaucratic maze ended. The two teachers were able to purchase the tower for minimal cost.

Today from the top of the former guard tower, no longer do searchlights peer into desolation and darkness. Instead of looking out at a no man's land scarred by dashed hopes of freedom, one sees twenty-year-old pine trees. They are part of a young forest of eighty thousand trees planted by Helga Garduhn and Marian Przybilla, along with their classroom students and now unified German Forest Youth club. The youth also have created a memory of victims of the wall. Plaques on four wooden pillars tell the stories of four young people who were shot close by. One pillar remembers eighteen-year-old Marienetta Jirkowsky, who wanted to escape to the West with her boyfriend. She almost made it, then they tripped an alarm wire, and the border guards opened fire. Marienetta died, her boyfriend managed to escape to the West. Today, the desolate strip of sand, weeds, and concrete—site of the occasional dead body of an East German desperate to scale the ominous barrier—has been transformed into the Berlin Wall Trail, and serves as a hiking and biking greenway.[3]

Przybilla remembers the helpfulness of the former guards. Instead of disposing of the saws and spades at the landfill, as they were instructed to do by the authorities, they brought the equipment to the environmental activists for their use. An old bunker and one of the four remaining guard towers—renamed the Conservation Tower—now house garden tools. Przybilla stands at the site and marvels: "It is remarkable what has happened here since—and all in our spare time." Despite his new position as a local lawmaker, he visits the tower daily, tending the garden and watering the trees.

Now in her seventies but younger than her years would suggest, Garduhn keeps up her struggle against the sand that was exposed over half a century ago when the fertile top soil was scraped away to discourage plants that might block the view from the guard tower. She crouches among the young trees and lays down a watering hose. And she guides visitors—adults and school children alike—on nature walks. They taste leaves, touch stems, and admire the different habitat types the children have cultivated around the tower. A small orchard, a pile of decaying wood, a wetland, an earthworm box, a honeybee colony, and a flower bed full of native plants—all tended by volunteers. Reflecting the cycle of life and death, a child reaches into the dead wood pile, pulls out a handful of earth and shouts: "That was once a tree!" Although Garduhn and Przybilla have received prestigious government awards for their work, it is these everyday fruits of their labor that reward them the most.

# 6

## Stewardship, Health, and Well-Being

*Civic ecology practices foster well-being.*

When they see something grow nice, they feel happy, like a flower. It helps in emotional healing. It's a place where people come and sit down, associate with each other. They come and pick it and cook it that morning like in Cambodia. They don't like to pick the bitter melon plants. They want to show them off to others. They are proud when people from the city come and take pictures. They want to make Lowell beautiful. They want to create a nice neighborhood.

—Saody Ouch, talking about Cambodian and Laotian refugee community gardeners in Lowell, Massachusetts[1]

A garden was one of the few things in prison that one could control.
—Nelson Mandela[2]

An iridescent metallic green beetle is killing ash trees across the Midwest—over 100 million trees since it first invaded the United States in 2002. And it is moving east, south, north, and west, killing ash trees along the way. Not long ago, ash was a common tree lining city streets, and in parks, forests, and swamps. But today, ash trees have all but disappeared from areas infested by the ash borer.

Yet somehow, the iridescent insect is doing more than killing trees. In the 244 counties where trees have been killed, over twenty thousand extra human deaths related to respiratory and cardiovascular disease had occurred just five years after the beetle invaded.[3] Why is the mere presence of the emerald ash borer associated with nearly seven additional deaths per every one hundred thousand adults each year? And why is the death rate higher in wealthier neighborhoods that previously had more ash trees? Although the increase in respiratory disease might be caused by poorer air quality as trees that once filtered out pollutants perish, the increase in cardiac deaths is harder to explain.

Why, when ash trees die, do people also succumb? We don't have a definitive answer. But we do have research showing how our health, and even the well-being of our communities, is dependent on nature. One study comes from Philadelphia where, for years, the Pennsylvania Horticultural Society has worked with community and city partners to transform blighted vacant lots into green spaces with grass and trees. But because the Horticulture Society has been more active in some neighborhoods than in others, the density of greened and still barren vacant lots varies depending on where you are in the city. In the parts of the city with greened lots, people feel safer and get more exercise. And gun violence is down near the greened lots.[4] The study of safer neighborhoods in Philadelphia is just one piece of the puzzle explaining how having nature nearby—even tiny greened lots in cities—makes us and our communities healthier.

The idea that nature benefits city residents is not new. Alarmed by the debauchery and disease precipitated by the rise of teeming industrial cities, nineteenth-century reformists proposed utopian rural farming communities and allotment gardening as a remedy.[5] Politicians also joined the chorus of urban greening crusaders. In 1850, both New York mayoral candidates called for a large park to counteract the depravity and decadence beleaguering the burgeoning metropolis. The result was Central Park, whose designers Frederick Law Olmsted and Calvert Vaux incorporated numerous natural features to secure not only "pure and wholesome air, to act through the lungs," but also as an "antithesis of objects of vision to those of the streets and houses." The views in Olmsted and Vaux's park would counteract the city's ills by acting "remedially, by impressions on the mind and suggestions to the imagination."[6]

It would be more than a century after the creation of Central Park that research, like the Philadelphia study of greened lots, began to confirm what reformists, politicians, and landscape designers knew intuitively. In a pioneering study published in 1984, scientists measured recovery times of patients who had just had their gall bladders removed, and who were randomly assigned to rooms with views of either trees or a brick wall. Patients who could see "green" regained their health more quickly.[7] Since then hundreds if not thousands of studies have reported the impacts of "seeing green" on human health and well-being. From our ability to think clearly to our capacity to recover from stress, from experiencing happiness to succeeding in school, from feeling generous to feeling good about ourselves, and from our spiritual sustenance to healthy interactions with neighbors and children, nature plays a role.[8] These benefits occur not just as we go about our everyday lives but also during periods of stress like when we are sick, during war, and after major disasters.

But seeing green is only part of the practice of civic ecology. Taking stewardship action—planting, pulling weeds, cultivating nature and community—are foundational to any definition of civic ecology. Meaningful action has important outcomes for our mental and physical health—in addition to the benefits of being around nature. Whereas witnessing destruction of favored places, or of trees dying and being cut down, can lead to feelings of helplessness and despair,[9] taking meaningful action leads to feelings of self-efficacy and hope. In this chapter, after a brief sojourn into the outcomes of seeing green, we take a longer journey to discover the health and community well-being benefits of civic ecology practices. We also ask: why does seeing, and "doing," green lead to so many positive outcomes for humans?

## Seeing Green

Imagine you are walking up to a tree, the trunk of which symbolizes seeing natural settings or "seeing green." Each of the branches of the tree represents one of the many benefits seeing green offers to people.

Although the benefits of seeing green are many, three big categories prevail: how being around plants and nature helps us feel restored—nature's therapeutic value;[10] how being in nature helps our brains function better—nature's cognitive benefits;[11] and how trees and green spaces create stronger ties among neighbors—nature's social benefits.[12]

Let's start with nature's social benefits first. Why would green spaces make neighbors feel more connected to each other? To answer this question, environmental psychologist Frances Kuo and her colleagues conducted a study of residents of the (much-studied and maligned) Robert Taylor Homes in Chicago.[13] Some apartment buildings had trees and lawns

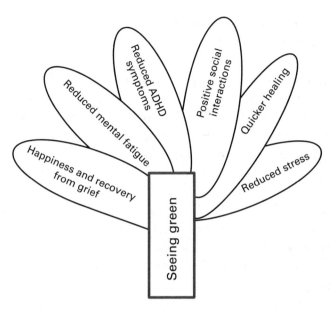

Well-being outcomes of views of green spaces or plants—of "seeing green."

nearby, whereas green spaces near other buildings had been paved over to cut down on dust and maintenance costs. It turned out that, compared to residents of buildings with no green spaces, residents of the buildings with nearby nature reported more social activities, knew more neighbors, had stronger feelings of belonging, and felt their neighbors were more likely to support and help them. The researchers tested whether these results—what they called neighborhood social ties—might be due to something about how nature affects people's brains. If being around trees meant people were less stressed, happier, or less mentally fatigued, would that lead them to forge stronger ties? It turns out that none of these psychological explanations accounted for why people in the greener apartment complexes forged stronger social bonds. The explanation was less complex. Simply greater use of the common spaces with trees and lawns, and less frequent use of barren common spaces, explained the stronger connections among those living in apartments with green spaces.[14]

Although reduced stress and fatigue did not account for the stronger bonds among residents at the Robert Taylor Homes, these cognitive benefits[15] are nonetheless important outcomes of being near nature. So are emotional or therapeutic benefits such as happiness[16] and recovery from grief.[17] Environmental psychologists (and spouses) Rachel and Stephen Kaplan have pondered the question of why nature might have these positive therapeutic and cognitive outcomes. They think it has to do with how we focus our attention. Humans spend a lot of time concentrating hard—for example when we read a difficult passage in a book, write an essay, deal with pain, or try to figure out the best way to approach a personnel issue at work. This type of concentration is called directed attention—and it's fatiguing, even stressful. When people tire of concentrating hard, they seek relief. Checking Facebook, chomping on chips, or taking a walk down the hall to the bathroom provides limited relief. But the Kaplans' research says: try something else. Put simply, take a break in nearby nature. This is because viewing or being in nature provides a powerful means to rest our minds and restore our attention. Nature—a brackish breeze off the ocean, a fledging red-tailed hawk squawking anxiously from a rooftop, the buzz of a flitting hummingbird, water bubbling up in an eddy, a squirrel scampering across snow, or wind blowing through tree tops—is intriguing, even fascinating. Switching our attention to nature's sights, sounds, and smells that intrigue or fascinate us seems effortless. It enables us to recover from the mental fatigue of concentrating and of dealing with difficult problems.[18]

The Kaplans' Attention Restoration Theory sounds good, but is there any actual evidence? For this we turn to the Japanese practice of forest bathing. The term forest bathing, or "Shinrin-yoku" in Japanese, was coined by the Japanese Ministry of Agriculture, Forestry, and Fisheries in 1982. It means "making contact with and taking in the atmosphere of the forest."[19] By 2004, Japanese scientists were so intrigued by the idea of forest bathing that they inaugurated the Japanese Association of Therapeutic Effects of Forests. And in 2005, the association undertook a study of Shinrin-yoku. The scientists asked: Do physiological markers of stress change after viewing or taking a walk in a forest? Results: forest bathers (those taking a walk in or viewing the forest) lowered their blood pressure, lowered their pulse, and lowered their levels of cortisol (the so-called stress hormone). This and other studies of nature and human physiology provide biological support for the Kaplans' psychological studies of nature and stress reduction. And the results may help explain some of the therapeutic benefits of seeing nature, such as quicker recovery times in hospitals, or even emotional restoration.[20]

Stress reduced, attention restored, and now we can go back to our desks. But our jobs can still be boring, perhaps even meaningless at times. Restoring our concentration and our ability to engage in stressful work is like weaving a beautiful tapestry but leaving it unfinished. We also seek meaning playing out on the tapestry of our lives.

## Doing Green

Civic ecology practice is not just about seeing green. Fundamentally it is about stewardship—doing green. It's about caring for green spaces and for the people who use them. It's about meaningful action. What benefits ensue from not just seeing, but also caring for nature?

Standing adjacent to the "seeing green tree," we can visualize a tree that depicts the added benefits of caring for or *doing* green. Augmenting the multiple social, cognitive, and therapeutic benefits of seeing green, we have three new branches: legacy, self-efficacy, and empowerment.[21]

### Legacy

Humans are not only motivated by status or opportunities to make money. We also share a desire to contribute something to the world. As we get older and reflect on our encroaching mortality, this desire acquires greater significance and even a sense of urgency. The psychologist Erik Erikson and colleagues have captured the notion of wanting to leave a positive legacy in a life stage they call "generativity."[22] Although awkward to pronounce, the term "generativity" takes on multiple meanings when applied to stewardship. It refers to a time in life when humans strive to leave a positive legacy for their children and future generations. One way to leave this legacy is through restoring—or regenerating—the land. When older generations nurture younger generations in learning how to regenerate the land, their legacy is strengthened. In a nutshell, through becoming environmental stewards, older people can generate something of lasting value, show care for younger generations, and in the process feel that their life will have meaning after they are gone. In so doing, older people make a "generative" contribution—at the same time contributing to their psychological well-being in ways not possible through simply seeing green.[23]

An older volunteer with an urban conservation group in Australia wondered out loud about his legacy—his generative stewardship experience:

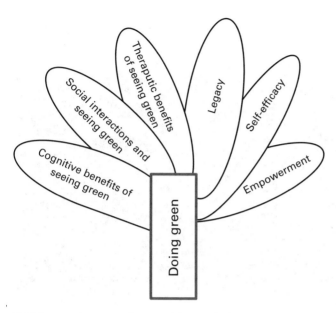

Well-being outcomes of stewardship—of "doing green."

You feel as though you have to be in there for the long-term . . . even though there is immediate gratification . . . I'm thinking of hundreds of years down the track . . . I look at a tree which I know, has a lifespan of over 500 years, and I'll try and picture it . . . I try and wonder what sort of people will be around and if our birds and animals we are trying to keep from going locally extinct, will be still around . . . I can't think of anything which is more long term.[24]

### Self-Efficacy

As their capacities decline further, older people may feel they are no longer able to do anything meaningful. But all of us at times are prone to feelings of uselessness. Hopefully, such feelings alternate with feeling that our actions are important. Sometimes people—young and old—have an unshakable belief in their ability to take meaningful action even in the face of overwhelming obstacles. Self-efficacy is the term used to describe a belief in our ability to take meaningful action and to exert control over events that impact our lives.[25]

Self-efficacy comes from seeing the difference that one's own efforts have made.[26] Studying hard and getting a good grade on a grueling test demonstrates to a student that putting in the effort leads to positive results. Planting a tree and watching it grow allows tree stewards to see

Volunteers—young and old—participate in a Brooklyn Bridge Park Conservancy landscaping day. Leaving a legacy for the Earth and its future generations is one of the well-being outcomes of "doing green." Photo credit: Alex Russ, Civic Ecology Lab, Cornell University

how their efforts make a difference. According to Stanford psychologist Albert Bandura, people "who have a high sense of efficacy visualize success scenarios that provide positive guides and supports for performance. Those who doubt their efficacy visualize failure scenarios and dwell on the many things that can go wrong. It is difficult to achieve much while fighting self-doubt."[27]

When experiencing threatening or difficult situations, some people are particularly subject to feelings of doubt and helplessness. The reactions may even be physiological—but the opposite of the physiological reactions of the Japanese forest bathers. According to Bandura, "When people try to cope with threats for which they distrust their efficacy, their stress mounts, their heart rate accelerates, their blood pressure rises, they activate stress-related hormones, and they suffer a decline in immune function."[28] The good news is that people who believe they can exert control over a situation experience less stress, less anxiety, and are less likely to conjure up disturbing thoughts. In this way, self-efficacy is vital to our ability to persist and recover when faced with adversity and even trauma.[29] It is also vital to our everyday health and well-being.[30]

Mastery experiences—proving to oneself that one's actions have a positive outcome—lead to feelings of self-efficacy.[31] When people share in an experience and some are more skilled than others, the more skilled guide the novices in mastering particular skills. Such "guided mastery" is one of the principal means by which individuals achieve self-efficacy.[32] It is easy to imagine how working to restore a patch of prairie in Chicago or a section of stream in Seattle would foster self-efficacy—through learning from skilled members of the group and mastering the restoration protocols. Furthermore immigrants and others who may have limited experience participating in civil society can learn from skilled community greeners about collective decision making and advocacy. Thus, stream restoration, block beautification, and other greening activities can serve as schools for civic participation, extending their impacts beyond individuals to broader community well-being.[33]

### Empowerment

Gardening Leave is a horticultural therapy program for war veterans in Great Britain.[34] Because the therapy includes active greening, it goes beyond simply quickening healing times by seeing green, as we saw in Ulrich's study of gallbladder patients looking out hospital windows. It links to a branch of community psychology that focuses on empowerment, or seeking to enhance the possibilities for people to control their own lives.[35] The notion of empowerment steers us away from viewing people as victims, people in need, and people requiring outside experts to cure them. Instead it guides us toward seeing people as playing active roles in leading healthy lives and being resilient in the face of obstacles. The psychologist Julian Rappaport calls for a shift in our thinking away from people as in need of our help, if our aim is to enhance the possibilities for individuals to take control over their own lives. "We will confront the paradox that even the people most incompetent, in need, and apparently unable to function, require, just as you and I do, more rather than less control over their own lives; and that fostering more control does not necessarily mean ignoring them."[36] In short, empowered people have power—to determine their own future. But in so doing they recognize that power can be shared and even expanded by people working together.[37]

Community gardening and greening has been shown to empower women in homeless shelters in California,[38] and residents in industrial U.S. cities.[39] But the starkest depictions of how people who have come to feel disempowered gain control over their own lives through civic ecology practices comes from the desperately poor and violent townships near Cape Town, South Africa.[40]

Immediately following the downfall of apartheid, women in townships had little control over their lives. Their husbands and boyfriends wielded power, not infrequently becoming violent when the women attempted, even through seemingly minor acts, to exert authority. In one instance, a woman was beaten by her boyfriend when she prepared *imifino*, a mixture of ground maize and a green vegetable, which her boyfriend did not deem fit for consumption by men. Following his tradition of Xhosa culture, the boyfriend believed that eating *imifino* would make him weak and tame, like women. But perhaps more threatening was the fact that his girlfriend had grown greens for the dish in her own garden plot. Maybe the boyfriend realized that even such a small act as gardening was an act of defiance—of empowerment. Gardening was beginning to give his partner and the other women control over their lives—by growing food for themselves and their families in a community where food was difficult to come by.

Often the women in the South African township gardened communally. Like the residents of Chicago's Robert Taylor Homes who forged stronger social ties by spending time in green spaces, the South African township women came to know each other, only this time through active gardening or "doing" green. The social bonds they developed enabled them to go beyond control over their individual lives. They began to exert a collective empowerment. At first the women were simply empowered to dare to talk among themselves about issues such as gender relations. But working with an NGO, a group of women gardeners began to get involved in local social and political action. In a triumphant albeit tragic event, the women demonstrated their newfound empowerment:

One of Nonyembesi's neighbours had gone out to a local spaza[41] the previous day to fetch some soap so that she could do her washing the next day when the weather looked set to be fine and dry. She left her 11-year-old deaf and dumb daughter at home with her husband, the girl's father. When she returned to the hokkie[42] she found her husband on top of his daughter, raping her. She rushed to Nonyembesi, who quickly got all the women of the street gardening group together. They returned to the hokkie to rescue the child and saw that the father was arrested and placed in jail. This in itself was a great achievement, as rape in the townships is rarely reported to the police and even more seldom are charges brought against men. The women's good work went further still, for while they were arranging for the man to be detained, some of the men in the street discovered what had happened and a group of them formed a lynch mob. By ensuring the arrival of the police, the women managed to avert a second tragedy. The women were even able to contact an NGO health worker who helped them to arrange counselling for the deaf child at Wynberg Hospital with a trained sign councilor. The next day, the women were all to be found in their gardens.[43]

The women gardeners living in Cape Town townships started out by taking control of small things, like the green spaces they cultivated and the food consumed in their households. From there, they became empowered to take on the ugly reality of violence in their community. The ties they developed through community gardening were instrumental to this outcome. Similar to how social capital is both developed through and enables further action, the gardening empowered the women, and the newly empowered women expanded their capacity for community action.[44]

Legacy, self-efficacy, and empowerment are some of the well-being outcomes of civic ecology practices. Civic ecology practices also provide fertile ground for spawning additional outcomes related to well-being, like expressing one's love of life and love of place, building social capital, and learning. But since these are covered in other chapters, we return now to the killer beetles.

### Killing Trees, Killing People

We have learned about the positive outcomes of seeing and doing green. How might this knowledge help us to understand the increase in heart disease-related deaths among residents of neighborhoods that have lost trees to the iridescent green beetle?

The authors of the emerald ash borer study do not offer an explanation for why the dead trees were associated with the increased human deaths due to cardiac disease. But here we can put forward a hypothesis. We start with the large body of research on the health impacts of seeing green, focusing on healing and reducing stress. The lost trees meant lost opportunities for alleviating stress through viewing and being in nature. As we broaden our purview to account for the impacts of doing green, a gaping hole in our tapestry appears. Although research is under way to find a means to control the emerald ash borer, to date there is little that can be done to save the ash other than expensive pesticide treatments of individual trees. In other words, people living in areas invaded by the emerald ash borer sat by and watched as trees outside their windows, and in places where they enjoyed hiking, biking, or simply visiting with friends under the shade of an ash tree, succumbed to the beetle. Likely they felt helpless to do anything about it. Such feelings of loss, coupled with lack of possibilities for meaningful action to save their trees, may have contributed to feelings of lack of control over their lives. Feelings of hopelessness and helplessness could have contributed, along with greater stress associated with less green space, to the increased deaths in Midwestern counties where ash trees have died.[45]

In the case of the emerald ash borer, civic ecology practices will not bring back the ash trees—at least not in the near future.[46] There is no denying that the demise of the ash trees is a profound loss. Yet, people and communities are resilient—they want to leave a positive legacy, to feel as if their efforts make a difference, to take control of their lives, and to live in healthier communities. And they will act on these feelings. Already in New York State, where the emerald ash borer is currently taking its toll, groups of volunteers and staff from government and conservation nonprofits are joining together to form local task forces to deal with the emerald ash borer invasion.[47] Their work entails planning for removal of the dead trees, spraying insecticides to save prized trees, and educating homeowners and communities about means to mitigate the economic impact. Concerned citizens also are sounding the alert when ash borers first invade a new area, by monitoring the activity of woodpeckers that feast on the invading insects. And task force volunteers are collecting ash seeds in the hope that trees will be replanted once scientists find a control for the destructive beetle. Collecting ash seeds, monitoring the infestation, and planting new trees of a different species—these are the small acts that help to provide a legacy, feelings of self-efficacy and empowerment, and hope. And these are the small acts that sow the seeds of an altered environment—but nonetheless an environment where people will continue to reap the benefits of seeing and doing green.

# Steward Story: Ben Haberthur

From as far back as he can remember Ben Haberthur heard stories about war. His mother and father served full careers in the U.S. Navy, his beloved stepfather served in the Marines and Navy, his twin sister is in the Air Force, and his youngest sister received an Army commission from West Point, not to mention the various military services of grandfathers, aunts, and uncles. So for Mr. Haberthur to join the Marines at the age of nineteen was part of a family tradition of military service.

In Iraq, Mr. Haberthur saw action as a machine gunner. His valor on the battlefield was recognized by a Combat Action Ribbon, two Marine Corps Reserve Medals, and a Presidential Unit Citation. But he also witnessed the environmental ruin brought about by Saddam Hussein. To retaliate against the Marsh Arabs who did not lend him their support, Hussein drained thousands of acres of Iraq's marshlands, and for the time his regime was in power, obliterated the Marsh Arabs' sustainable livelihoods and traditions.

After his military service, Mr. Haberthur earned a degree in environmental science from California State University. Now back home, Haberthur's encounters with nature led him to reflect on his future. "I returned to school, anxious to get on with my life, and I discovered, while exploring the coastal areas of California, nature provided a peaceful and calming alternative to the stresses of my former military life."[1] Military service and restorative nature were once again to intersect when Mr. Haberthur returned to his native Illinois, and found himself working for the Forest Preserve District of Kane County. There he discovered the Dick Young Forest Preserve, named after local conservation legend and World War II veteran, Dick Young, who served in the legendary battle at Iwo Jima.[2] According to Haberthur, Dick Young "was able to overcome all he saw on Iwo Jima to become a leader in the fight to save our region's natural areas. He embodied the belief that a country worth protecting is worth preserving."[3]

Combining his passion for veterans and for nature, Mr. Haberthur established the Veterans Conservation Corps of Chicagoland. Today, he works with veterans desperately seeking to escape from their tormenting memories, and to find a way to evade alcohol abuse, violent behavior, depression, and all too often suicide. With support from the National Audubon Society's TogetherGreen Fellowship, Mr. Haberthur and fellow veterans plant trees, clear brush, and remove invasive species, and have cleaned up a marsh wetland on the Dick Young Forest Preserve. And they have brought their families to the preserve for a cookout—to celebrate their accomplishments and just being alive and together.[4] In helping to bring back native wildlife and plants and to heal nature, Mr. Haberthur and the veterans are "bringing back" and healing themselves.

# 7

## Learning like Bees

*Civic ecology practices provide opportunities for learning.*

### The Bees and the Band

Contrary to popular opinion, a colony of honeybees, according to Cornell University biologist Thomas Seeley, is not governed by a benevolent female dictator. The matriarch of the hive—the bee Seeley refers to as "Her Majesty the Queen"—does not direct everyday activity, and she is not the "Royal Decider" for all the colony's affairs. Instead, based on years of research compiled into his book *Honeybee Democracy*, Seeley asserts that the queen's function is to lay fifteen hundred or so eggs per summer day, and thus she more aptly is named the "Royal Ovipositor."[1]

Having relegated the queen to a purely reproductive rather than a thinking role, Seeley goes on to describe how the worker bees in a hive, lacking an overseer, create a "democratic" society whose collective abilities far transcend those of the individual bees. Further, a swarm of honeybees, similar to the cells in a human brain, demonstrates a collective intelligence. This collective intelligence is used each year when the colony faces a critical dilemma.[2]

Every spring honeybees face a life or death decision—the results of which determine their colony's survival. The bees' nesting site—a cavity in a tree or a wooden-frame hive—becomes overcrowded and they are forced to seek out a new home. To find a new nest site, they need information about the environment: What site best supplies pollen and nectar, protection from predators, and space for their young?

How does a colony of honeybees make this life or death decision? Out of a swarm of ten thousand worker bees and one queen, several hundred nest scouts take flight to gather information across a swath of seventy square kilometers around their old hive. The scouts then return to the swarm and share what they have learned with their fellow colony mates through an elaborate ritual of waggle dances. Each move, intently watched by the other bees, communicates information about the location and quality of potential new homes. The swarm makes a collective decision based on the information that is presented, and flies off to colonize its new tree cavity or hive.

Thus, faced with a life or death decision about their future, bees need to gather information—to learn—about their environment. The outcome of their learning—which nesting site is most promising—is not known in advance. If each bee were to learn about its environment in isolation and not share information with the colony, the colony would perish. Instead honeybees have evolved a social or "democratic" system to create new knowledge and to take action to ensure their collective future. Learning in civic ecology practices also is "democratic"—people with different types of knowledge share their expertise, and together they take collective action to ensure their own and their community's well-being.

A jazz band on first take seems to have little to do with a colony of honeybees. But like the decision honeybees make about where to construct their new nest, the outcome of an improvisational jazz session is not known at the start. And like the honeybees, the musicians interact with and learn from each other. Then, something magical emerges from the improvisational back and forth—the interactive "dance" of the musicians. The magic that emerges is music.

Using the analogy of a jazz band as his starting point, environmental education scholar Arjen Wals and colleagues have written about the *Acoustics of Social Learning*.[3] They describe social learning as bringing together people of different backgrounds to create an "ensemble of perspectives, knowledge and experiences." The goal of Wals's social learning—not so unlike the goal of civic ecology practices (or of a colony of honeybees)—is collective action toward a more sustainable future. But the outcome of the learning experience—the actual action steps that will be taken—is not determined a priori, just as when members of a jazz band bring together their different instruments and talents, and improvise to create music. Like the honeybees and civic ecology stewards, musicians in Wals's "jazz band" are social learners setting out to solve an environmental problem. They seek out and share information, and then collectively commit to a course of action based on the information gathered.

Wals and his colleagues outline five principles for social learning toward sustainability,[4] which we paraphrase here giving examples from Seeley's democratic honeybees and our civic ecology practices.

1. We learn from each other.

The colony learns from the nest scouts.

Civic ecology stewards learn from each other.

2. We learn more in groups of people who don't all think alike.

Each scout bee presents a different solution to the nesting dilemma.

Civic ecology stewards come from diverse backgrounds and bring different perspectives to the table.

3. Trust and social cohesion are essential building blocks in the process of learning from people who hold different views.

Members of the bee colony, according to Seeley, "respectfully" listen and give consideration to each nest scout's solution.

Civic ecology stewards form social ties and trust through working together, which provides a foundation for learning from each other.

4. Social learning is a process of collectively coming to understand a situation.

Through watching the waggle dances and each scout bee's moves, the colony of bees develops a common understanding about the optimal solution.

Through considering their fellow stewards' points of view, civic ecology stewards attempt to develop a mutual understanding of how to address problems they face.

5. Social learners help to create the learning process and the solutions to the dilemmas they face, and thus are more likely than passive learners to follow up with action.

Rather than the solution being dictated to a passive colony by Her Majesty the Queen, the bees take collective action when they fly off to a new home.

Civic ecology stewards not only take action to green their neighborhoods. They also advocate and work collectively for greening- and community-friendly policies.

The first four principles build toward the fifth, in which the social nature of learning is linked to collective action. Social learning thus brings together people with different perspectives and experiences to try to find solutions to environmental and other dilemmas.[5] One such dilemma—the dilemma that is addressed by civic ecology practices—is how to transform broken places housing broken communities into something more nourishing for people and the environment. And social learning plays a big part in how we address this dilemma.

### Learning from Interactions with People

Wals's jazz band offers a contrast to how we usually think about learning. Social learning is less about a student acquiring knowledge transmitted by a teacher in a classroom (or dictated by a queen bee). And it is less about the wealth of content that constitutes environmental or scientific literacy and is evaluated on standardized tests. Social learning, in contrast, focuses on learning through interactions with other people—and what one learns is not predetermined. Like the honeybees and the jazz band, the learning emerges from the interaction.

In the jazz band each person, regardless of whether he is an expert or a novice, brings his own instrument. Civic ecology practices are similar—people with different kinds of knowledge come together to share their expertise. Ben Haberthur is an Iraq War veteran who works with other returning warriors to restore the native forests near Chicago. Each veteran brings something unique to the restoration effort—one knows how to use a chainsaw, another has experienced the power of nature to heal psychological wounds, and a third is an expert on which trees to cut and which ones to protect. Some may have grown up in the countryside, and have memories of how fire was used to control unwanted vegetation. Including everyone's knowledge, experience, and memories is important because no single individual can profess to have all the information needed to carry out a complex task, or to make a forest management decision.[6] In the case of Ben Haberthur's veterans and the forest project, the end result of learning from each other, and of their collective wisdom, is action to restore a healthy woodland—and hopefully, a community of veterans with their emotional war wounds at least partly assuaged through being in and restoring nature.

Whereas in the veterans' forest restoration project each participant brings his own skills and knowledge, most likely not all the participants' levels of expertise are equal. Novice chainsaw operators work alongside experienced old-timers. The way in which the novices learn is not unlike the way aspiring tradesmen during medieval times learned through be-coming initiated into a guild. The novice began his learning process by observing the more experienced tailor or farrier. After a period of watch-ing from the sidelines, the novice ventured to copy the master's simple stitching or fitting.[7] He likely benefited from scaffolding—a series of higher-level challenges presented by the master. As the novice interacted with the expert—through imitating, asking questions, trying out new skills, and receiving feedback on his performance—he gained expertise. The novice who persisted eventually became a full member of the tailor or farrier guild. "Communities of practice" is the term coined by education scholar Etienne Wenger and anthropologist Jean Lave for modern-day social learning environments that resemble medieval guilds in their use of scaffolding and other interactions among novices and masters.[8]

One can imagine any number of civic ecology practices that provide opportunities for learning defined as increasingly more skilled participa-tion in a community of practice. For example, Haberthur's veterans group is part of Chicago Wilderness—a longstanding network of scientists, non-profit organizations, government agencies, and volunteers—who form a

Students from Satellite Academy High School listen to and interact with each other and their urban environment at the Eagle Street Rooftop Farm in New York City. Photo credit: Alex Russ, Civic Ecology Lab, Cornell University

community of practice built around native forest and prairie restoration in the Chicago region. It includes among its many activities opportunities for students to collect and plant native seeds with experienced adults, and for adults to learn about controlled burning to rid a site of non-native, invasive species.[9] Similarly, the Garden Mosaics education program provides opportunities for youth to work alongside adult community gardeners, who have created a community of practice around converting urban vacant lots into a mosaic of green spaces that become places for meeting up with neighbors, growing food, savoring feasts, performing plays, creating memories—and learning.[10]

However, neither the jazz band nor the medieval guild paints a complete picture of learning in civic ecology settings. Social learning focuses first and foremost on interactions with other people. But remember the bees? They were not only interacting with each other but also venturing out and collecting information about the environment. Learning occurred not just through interactions with their fellow bees, but also through interacting with the air, soot, pollen, and dirt that constitute their surroundings. How do similar interactions with the environment foster learning among humans?

### Learning from Interactions with the Environment

Let's jump from the bees to the Bronx—and to the youth organization Rocking the Boat. Rocking the Boat affords youth in their programs, as well as families who live nearby, multiple opportunities for learning. It houses an immaculate woodshop where young people learn to construct quality wooden boats—replicas of the historic rowboats that once plied New York's harbors. Moving from the sweet smell of fresh-cut wood inside, outside to the sulfurous estuary nearby, youth at Rocking the Boat help restore the once-abundant Eastern oyster to local waters. Like the bees sampling possible nest sites, they not only smell the Bronx River, but also test its water quality to determine the river's suitability as a home for oysters. Finding the habitat of sufficient quality, they launch their wooden rowboats laden with oyster spat and oyster cages into the Bronx River, and lower the baby bivalves and their artificial homes into the murky waters. Through interacting with more knowledgeable leaders, and with scientists studying oysters, the teens learn about how oysters filter pollutants from the water and form reefs. They learn that in the past abundant oysters provided ecosystem services like water filtration and shoreline protection, and that if the bivalve populations increase through the efforts of oyster gardeners like themselves, oysters may once again provide these services. In other activities, the youth become "citizen scientists" collecting data on wading shorebirds, and sending it to professional scientists at the Cornell Laboratory of Ornithology. Not unlike the information collected by the nest scout bees, the data collected by these human "scouts" contribute to decisions about how best to manage the shoreline habitat for wading birds. And on summer Saturdays, the more experienced and knowledgeable youth take neighborhood residents out on the river for community rowing days, helping their passengers learn about the river and its inhabitants.[11] At Rocking the Boat, "Kids build boats and boats build kids."[12]

Throughout their learning activities, youth at Rocking the Boat interact with each other, their leaders, scientists, and families living nearby. At the same time, they interact with the wood and the water, the oysters and the birds, their Secchi disks for measuring water turbidity, and their oars for getting where they need to go. The people, the animals and tools, and the wood and the water all afford opportunities for the young people to learn.[13] They learn about the properties of wood and the habits of shorebirds, how to build trust and work together, and how to participate in broader communities of citizens and scientists who collect data and make conservation decisions.

Whereas a young person at Rocking the Boat is surrounded by a suite of living and nonliving things, including people, ideas, water samplers, estuary muck, egrets, and oysters, it is the meaningful interaction between the young person and what surrounds her that enables learning. If she launches out on a rowboat intent on becoming an oyster gardener and the cages are too heavy for her to lift, or if she can't understand the scientific language of the other oyster gardeners, then she will have difficulty interacting with the oyster cage and the people; they will lose meaning relative to her learning. At the same time, if the young person lacks curiosity, a work ethic, basic knowledge of science, and a willingness to listen and to be serious about the activities, she will not be able to interact productively with the objects and organisms around her; she will fail to learn. Thus we need to consider not just how living and nonliving elements of the environment—and activities like restoring oysters or collecting data on shorebirds—make possible learning. We also need to consider the match—and importantly the interactions—between the learner's skills and other competencies, and the environment around her.[14]

In short, learning is a two-way street—a dance back and forth between the learner and the other people and living and nonliving inhabitants of her environment. It is about the interaction between the elements of the learners' environment and what the learner brings to the table.[15] And if in fact learning is an interaction between the learner and aspects of the system that surrounds her, one might ask not only how the learner changes, but also what happens to the larger social-ecological system with which she interacts and of which she is a part.

### An Ecology of Learning

When we talk about learning, we nearly always focus on the individual students. Occasionally we might consider whether a community or even an organization learns. Might a neighborhood, a landscape, or even a complex social-ecological system like a city "learn" too?

A volunteer in New Orleans who learns to plant and care for trees experiences changes—maybe even transformations. He can now do something he couldn't do before. He has changed from simply being a resident of a neighborhood to becoming a member of a tree stewardship community of practice. But as he walks through his neighborhood, he can see that the neighborhood also has changed. City blocks covered by asphalt and concrete, or post-Katrina wastelands of mud and rubble, begin to sprout shoots and green leaves. Soon the trees are providing roosting sites for birds and shade where mothers linger with their toddlers learning to walk. A neighborhood that was silent now springs to life with the chattering of starlings and babbling of toddlers. As the learner transforms, so too do the local systems of which she is a part.[16] If people in other neighborhoods learn about the tree planting practice and the changes it has wrought, they may want to do something similar and the activities in the first neighborhood can bubble up and instigate changes in other neighborhoods and eventually the city.

Zooming back down to the learning activity, we can see another type of transformation—that of the activity itself. This time we consider an activity called "reef balling." To picture a reef ball, imagine a concrete igloo with Swiss cheese-like holes that provide places for clams and corals to latch onto and for fish to hide from predators. Reef balls come in different sizes, shapes, and colors, and can be designed, constructed, and deployed by Boy Scouts, angler associations, and other environmental and community groups.[17]

A reef ball project might start out with the goal of reestablishing a sport fishery—such as a project along the Rhode Island coast to enhance the number of tautog, black sea bass, scup, and cunner fish.[18] In partnership with scientists, anglers monitor the fish populations, and adjust the reef ball design to encourage more fish. As they learn about other reef ball projects from Connecticut to the Caribbean Islands, they may decide to add near-shore reef balls to their activity and seed them with clams and oysters. And they may also decide to join forces with Eternal Reefs, Inc., to promote use of reef balls as a means of celebrating the legacy of loved ones who loved the sea, by encasing their cremated remains in the concrete igloos and placing them among the structures that are rebuilding coastal reefs.[19] But upon viewing firsthand their favorite beaches washed away in a hurricane, or learning about how the last remaining Victorian house on what once was a thriving island community with stores, a school, and a baseball team is succumbing to the encroaching Chesapeake Bay waters,[20] the anglers and scientists may question their focus on restoring marine biodiversity and providing an outlet for grieving. While growing fish, clams, and oysters is important—they provide food and filter contaminants from the Bay—if we don't address sea level rise we may end up with worse scenarios than eutrophic estuaries and a short supply of fish. Seeing rising sea level as the "new normal" that must be addressed, the reef restorers may decide to transform their original activity—cultivating marine life using reef balls—into an expanded effort that encompasses not only designing and placing reef balls to maximize fish production, but also to protect shorelines.[21] The "reef ballers" might also expand their efforts to encompass new learning activities about adaptation to sea level rise, or attempts to influence local policymakers to incorporate mitigating and planning for climate change. The anglers' thinking has been transformed by their learning, and the anglers have transformed the original activity into something larger.

Harking back to notions of develop and deploy—or feedbacks—from our earlier discussions of social-ecological memories and social capital, civic ecology settings are characterized by ongoing and reciprocal interactions between the learner, the community, and other aspects of the environment. These interactions afford people with opportunities to build greener and more civic-minded cities. A city that is richer in nature and civil society activity, in turn, provides new opportunities for learning. In short, the individual learns as he brings about changes in the system that he is part of, and these changes enable further learning. This is one of a number of feedback scenarios in civic ecology practices that start with a small learning activity—like learning how to plant a tree or to fill the nooks and crannies of a reef ball with baby oysters. Describing the iterative feedback between learners and their environment, social learning scholar Claudia Pahl-Wostl remarks: "the learner is changing the environment, and these changes are affecting the learner."[22] At Rocking the Boat, remember, "kids build boats and boats build kids." In civic ecology practices, the *environment builds people and people build their environment.*

**What Do We Learn and Why Do We Care?**

It is also clear that complexity, uncertainty, risk, and necessity are ineluctable facets of a coevolutionary understanding that cannot be wished (or educated) away.[23]

Any environment is complex, uncertain, and poses risks to its inhabitants. And those inhabitants need to continuously learn in order to survive. Thomas Seeley's honeybees face a complex array of inputs (soot, pollen, rain, wind speed, angle of the sun). It is uncertain whether they will find a new home. Every honeybee faces the risk of being eaten by a bird, flattened against the windshield of a fast-moving Ferrari, or blasted by panicked picnickers squirting Raid Flying Insect Killer. And they all face the necessity of finding a new home.

Humans also face complexity, uncertainty, risk—and the necessity of finding a way out of the social and environmental dilemmas that confront us. It would be easier if there were a simple linear solution—like a teacher instructing us to add sugar to a petri dish to make yeast grow. Instead the social-ecological systems we live in are complicated and fraught with risk. Learning—which entails interactions leading to collective action—is necessary to deal with that complexity. So far we have talked about the process of learning—through interactions with people and the environment—and how people and the environment are transformed through these interactions. Here we ask: What is learned through civic ecology stewardship, and how does that address the complexity and uncertainty inherent to social-ecological systems?

Given how complex and uncertain social-ecological systems are, managers—whether they be professional foresters who control large tracts of timber, community gardeners in a small vacant lot, or volunteers restoring the Eastern oyster—continuously conduct small field experiments and adapt their management practices based on the results. Volunteer efforts to restore degraded prairie and forests around Chicago demonstrate how, through a series of experiments such as controlled burns to suppress invasive species, lay people and scientists refined their restoration protocols.[24] The leaders of the prairie and woodland restoration practice also experiment with ways to recruit and motivate the participants—will a cookout with hotdogs or a potluck dinner with an array of ethnic dishes make participants feel more connection to the group? Through their stewardship activities, participants learn about the outcome of changes in the plant varieties; of various watering, fertilizing, and fire regimes; and of different means of clearing vegetation or rewarding the efforts of volunteers. And this set of learning outcomes is important because it provides critical information that is used to improve the civic ecology practice, and to adapt it as the surrounding environment and the neighborhood inevitably change.

Learning that informs how best to adapt practices is necessary but not sufficient for management of a common resource like the Chicago prairies and woodlands. As we mentioned in chapter 3 on social capital, when multiple community members share a resource, several factors determine whether or not it is managed for the common good or becomes a tragedy of the commons. Learning about diverse perspectives and to trust others are two of the factors needed for successful management of a shared resource. A young woman from Brooklyn describes how she learned about diverse perspectives and trust through participating in an urban gardening and farm stand program:

I think my life is different cause, having to work with a bunch of kids and then . . . they always pair you up with a different person . . . a person of different gender or different background and . . . like I never used to like trust people and it's like, I always like to be by myself, wanted to do everything on my own, I didn't want no help, nothing. And then coming here and having to work with another person and they havin' to help me, shovel something or doing somethin' that had to be worked in a pair and then having to trust the person, that the person's gonna do it and I think . . . giving trust is easier for me to do now.[25]

Learning about the environment is also part of civic ecology practices. A second young woman from the Brooklyn gardening program learned about ecosystem services and how to share that knowledge with others:

Like how to talk to someone, when you cross the street, get your point across and they would know what you was talkin' about? So that helped me, you could start off the conversation like, "Oh, do you know that mostly all of our leaves go to the landfill in New York, New Jersey and all of that could be composted and put into nutritious dirt, for the soil and it's going right back into the earth? It's like a cycle process." And they're like, "Oh, that's true, . . . I just threw a bag of leaves in the garbage." I was like, "See, that could have been right here, you could even put it in a jar or something and bring it to us," and they be like, "Okay, so next time I'll rake my leaves and bring it." I was like, "Yes, that's a good idea . . . and. . . . Oh, you know when you cut your grass? That grass could go in compost too. You don't have to mow it on the streets and then it goes into ocean and then it just sits, stays there just like seaweed."[26]

And a young person from Rocking the Boat talked about how she learned new meanings about the Bronx as a place:

I have lived in a project housing near the Harlem River for at least 10 years now. . . . My apartment overlooks the Harlem River, but there is no access to that river. It's all fenced up and trains pass by, so you never can go to this river or at least you never thought you could. . . . When you go to school, they talk about different countries and the history of New York City. But they never talk about the environmental aspects of the city or its rivers and parks. Teachers don't tell you that hawks fly and catch some prey in the city.[27]

Jumping to her boating experience with Rocking the Boat, this same young woman describes a transformation in her thinking about the place where she lives: "When we were on Harlem River I looked to my right and I saw my house in the Bronx, my area, and my apartment. I was like, 'Am I on the river that I look at every day? I am on this river!'"[28]

And finally, we visit community gardener Willie Morgan who grows cotton in a tiny garden in Harlem. Willie Morgan comes from the Southern U.S. black slave and sharecropping tradition—a harsh way of life which he sought to escape by moving to New York as a young man. When asked why he grows cotton on a tiny plot in Harlem, Morgan explains: "Because I want the young folks growing up today to understand where they come from—to understand their roots."

Like the music that emerges from interactions of the musicians in the jazz band, learning in civic ecology practices can take on any number of forms. How to conduct small-scale experiments and adapt practices based on their results, how to build trust and honor diverse perspectives, how to explain ecosystem services, and how to connect to a river and to social-ecological memories—all are learned through interactions with people and with the environment in civic ecology practices. And collecting information, building trust, listening to other perspectives, understanding ecosystem processes, and connecting to the places where we live, all contribute to taking collective action to sustain a common good—and to our deeper learning and engagement in civic ecology practices.

# Steward Story: Andre Rivera

Andre Rivera grew up in an apartment in the Bronx's "toxic triangle."[1] Three major expressways—the Sheridan, Cross Bronx, and Bruckner— along with the toxic exhaust spewed out by their bumper-to-bumper cars and trucks, define the triangle. Also near Andre's apartment is the Bronx River, but he had never been there—that is, until he started getting into trouble in school and his teachers told him to go to an after-school program to improve his social skills. Andre's mother found one on their block—at a community organization called Youth Ministries for Peace and Justice (YMPJ).

Andre remembers the first time he got onto the Bronx River. He was in middle school, and together with the other YMPJ youth, he embarked in a canoe in the South Bronx near the old concrete plant. The water was frightening because he had never been out on the Bronx River. Nor had he ever seen his community from the river vantage point. As Andre and his YMPJ companions floated down the river through the South Bronx, he saw a floating rat and a lot of floating trash. This was Andre's first experience of the Bronx River not far from where he lived.

Like nearly a fifth of the people who live in the Bronx, Andre suffers from asthma. At YMPJ, he joined the asthma campaign—fighting against air pollution in his borough. He also started working with environmental educator Steve Oliveira. It was Oliveira who introduced Andre to canoeing, as well as to gardening and conducting fish studies in the Bronx River. He showed Andre "how to catch fish and what types of fish there were, what types of trees there were, how to plant a tree, how to plant native species, how to identify them and how to pull up weeds."[2]

Andre continued his work at YMPJ through high school, joining a campaign to transform the Sheridan Expressway into a more community-friendly space. While training other youth in the campaign, Andre also volunteered to help construct a Harvest Dome out of recycled materials. The dome was intended to be an educational display tracing the Bronx River watershed and its sources of pollution. On the day the Harvest Dome was to be floated from Hunt's Point in the South Bronx, to its new home at a nature center in Inwood Inlet in Harlem, Andre got sick. And it was windy and rainy. The other volunteers placed the dome on a pontoon made from several canoes strapped together, and embarked on their journey south. But the pontoon began taking on water and the wind carried the dome to the no man's land that is Riker's Island. Since you can't land on Riker's Island, the dome was lost. Andre is pretty sure that the Riker's Island prison guards hauled it up on the beach and destroyed it.

Andre continues to be drawn to the river where he was first exposed to nature. In addition to that first canoe trip, another memorable experience was trying to get to Goose Island. He and other environmentalists started out in four canoes and one speed boat, intending to pick up trash on the island. But the tide had gone out, and the canoes got stuck in the muck. Andre used the speed boat to rescue the others, now waist deep in muck. Despite this misadventure, Andre says, "I am a canoe person by nature, and I am on river by nature."

Today Andre continues his work as an environmental educator at the Bronx River Alliance. He educates school and community groups about the Bronx River and leads community canoe expeditions. The groups launch near where Andre had his first experience on the river—from an abandoned concrete batch mix plant that the Bronx River Alliance and YMPJ helped to transform to Concrete Plant Park. Thanks to the collaboration among community groups and government, today the park includes grassy play areas, trees, a reading circle, restored salt marshes, and a science teaching lab, scattered among relics of silos, hoppers, and conveyor structures that serve as a reminder of the park's industrial history.[3] Andre rides his bike through the park every day on his way to and from classes at Boricua College. He is about to receive his associates degree in liberal arts and plans to go on for his bachelors in human services. Then he wants to get a masters degree in public policy and political science. His goal is to be a leader for the Bronx community and for New York.

Despite the fact that many still view the Bronx as an industrial waste-land, Andre finds nature along the Bronx River. He uses nature as an escape from the stresses of life for a young man in the Bronx—to provide a sense of harmony. He also realizes that unless you live in the city, you probably won't see the nature within it. And even people living in the city don't take the time to see how nature has made its way into their neighborhoods.

But Andre believes that without connecting to nature, being able to accomplish things in life is difficult. So he has taken the time to see the beautiful river. And he has marveled at the beavers and mute swans re-turning to the river for the first time in years. And through nature, he has connected to his community and seen that everyone's struggle is parallel to his own. Like the beavers re-inhabiting the Bronx River, he is not alone.

# The Second Interlude

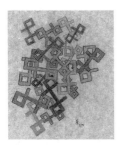

Art: Keith G. Tidball

We have now completed our "bricolage"—we have assembled the pieces of the civic ecology practices. Sense of community, social capital, and collective efficacy, social-ecological memories, places and faces, ecosystem services and human well-being, and finally learning about all these things—are pieced together in different ways. Often the piecing together is not planned out beforehand—the pieces "self-organize"—and the civic ecology practice emerges.

All along the process of bricolage, we have seen that as we put the pieces together, they interact with and change each other. And they create greater possibilities for action. Through planting a tree in New Orleans or cleaning up litter in Tehran, civic ecology practices provide shade and places for people to enjoy nature. These ecosystem services encourage people to spend time outside, and even to join in the tree planting and litter cleanup activities. And through learning about the healing power of nature and learning to trust each other as they restore bald eagles to the Anacostia River, youth are desperately trying to transform nature and their communities in Washington, DC, and to create new, healthier social-ecological memories they can draw on in the future. In short, as the pieces of civic ecology practices interact, their sum becomes greater than the parts.

Each individual civic ecology practice can be thought of as its own system—a system of processes, places, and faces, and their interactions and feedbacks. But what happens when civic ecology practices interact with people, organizations, and the environment outside the particular practice? We take up this question in the chapters in part III, on governance and on social-ecological systems.

# III

## Zooming Out: A Systems Perspective

# 8

# Governance

*Civic ecology practices start out as local, small-scale innovations and expand to encompass multiple partnerships.*

When you remember who you are, and you are connected, you build from the inside out.
—Alexie Torres-Fleming, Bronx native and founder of Youth Ministries for Peace and Justice[1]

## From the Center Out

A father and his young son stroll down the sidewalk on a stifling, summer day in Houston's Third Ward and come upon a small green oasis—the Alabama Community Garden. The casual strollers are drawn in by the aroma of chicken smoking on a grill next to the garden shed.[2] They gaze in wonder at the lushness of the tomatoes, beans, and mustard greens thriving in raised beds, and wander over to where children are intently watching birds dart in and out of wooden houses nailed onto the trunks of trees. The gardeners, sitting on a bench beneath the shade of a live oak, welcome the passersby into their oasis, to look around, to smell and to taste, to share stories, to escape from the traffic rumbling by on Alabama Street. The visitors become intrigued by the bounty the gardeners have cultivated, by the quiet sanctuary from the heat, clamor, and asphalt of the street, and by the opportunity to join in efforts that convey a renewed sense of civic life.

Not only the father and son, but also city government and nonprofits may take note of the Alabama Community Garden, sizing up its contributions to larger civic renewal and sustainability initiatives. A city assemblyman may stop by to learn how the gardeners make collective decisions about managing their common resource. And a board member of the nonprofit Urban Harvest might check on the amount of produce being grown with an eye to calculating the total contribution of community gardens to healthy food production in Houston.

Earlier we talked about how civic ecology practices help to create a sense of community and social capital among participants within the practice. But what happens when people who are not active gardeners in the Alabama Community Garden stop by to taste a tomato or to check out the possibilities for partnering with the garden to achieve broader goals? How might the influence of the garden begin to extend beyond its borders, and how might it be part of larger systems of governance?

To answer these questions, we can imagine a series of concentric rings with the garden and its active gardeners at the center. The next larger ring is the surrounding neighborhood—the dads with the toddlers who stop in to join the activities, as well as the housebound seniors living nearby who receive fresh produce delivered by the gardeners. And a still larger ring includes the other community gardens in Houston along with the network of the nonprofit organizations and government agencies with whom they interact—the "governance system."[3] For community gardens in Houston, the network or governance system includes a long list:

• *At the local level,* community gardening organizations such as: the City of Houston Community Garden Program within the Department of Health and Human Services and the eleven community gardens in its network, the City of Houston Parks and Recreation Department along with the five community gardens on its properties, and the nonprofit Urban Harvest which supports a network of over one hundred gardens in the Houston area.

• *At the national level,* community gardening organizations including: the U.S. Department of Agriculture Cooperative Extension Service and People's Garden program, as well as the nonprofit American Community Gardening Association.

• *At the local, national, and global level,* organizations and practices that focus on activities related to community gardening, like civic agriculture, education, and urban sustainability: farmers' markets, schools and youth programs, the city's sustainability office (Green Houston), and the Houston nonprofit Citizens' Environmental Coalition—as well as international groups like the Urban Biosphere Initiative (URBIS) network of cities working to enhance urban biodiversity.[4]

But all these organizations being involved in community gardening may seem like overkill. Why not just one central organization that supervises the network, formulates the policies, and enforces the rules? What's so important about a system with so many centers of activity—what Nobel laureate Elinor Ostrom has referred to as a "polycentric" governance system?[5]

### Multiple Centers—Polycentric Governance

The notion of governance offers an alternative to the more limited definition of formal government. It recognizes that multiple institutions and organizations influence policy. Governance institutions include not only city, county, state, and national governments, but also businesses, community groups, and small nonprofit organizations, as well as large national and international NGOs like The Nature Conservancy, and multilateral organizations such as the United Nations. Based on multiple studies conducted over many years on systems as different as policing and forest and water management, Elinor Ostrom concluded that, relative to top-down government, multiple layers of governance do not produce inefficiencies but rather enable societies to more effectively address complex challenges. But how, Ostrom asked, can overlapping systems be so effective?

Some of the answers to Ostrom's question can be found in the previous chapters of this book. And we have already given you a hint: polycentric governance systems allow priorities and policies to be set at multiple levels—ranging from the community garden, to neighborhoods and cities, to state, national, and international organizations. In this way, polycentric governance brings multiple voices to the table, and enables a diverse set of options that would not be possible if only the larger institutions were present. All voices have something to contribute, just like the diversity of organisms needed to provide ecosystem services. Civic ecology focuses on the contributions of the smaller players, but its practices interact and are networked with organizations at multiple levels.

Community gardens, such as the Garden of Happiness in the Bronx, provide opportunities for multiple voices to engage in discussion and decision making. In this way, they serve as training grounds for building the trust, experience in resource management, leadership, and other skills important to participation in larger polycentric governance systems. Photo credit: Cornell University Cooperative Extension-NYC

How might small, local practices be important in polycentric governance systems? In chapter 3, we talked about the importance of connections and trust in enabling management of shared resources. We also mentioned how community norms, in particular expectations for behaviors that are in the public interest, can lead to collective action in addressing crime, litter, pollution, and other local problems. Polycentric governance systems encompass local organizations—and it is at the local level that people develop the connections and trust needed to collectively manage shared resources. Similarly it is at the local level that adults tell kids to stop harassing passersby or to pick up their trash, or that neighbors transform a vacant lot into a community garden and that people demonstrate the collective efficacy critical to addressing neighborhood incivilities and crime. Topophilia, or attachment to place, also is mostly a local phenomenon and may motivate people to work together to improve their local environment and community.[6] Leadership as well begins locally.

In chapter 5, we explored how some civic ecology stewards measure the biological, physical, and cultural services they provide. Polycentric governance systems enable collecting and communicating this kind of information at the local level, which is critical to developing rules about how we manage a particular shared resource. And in chapter 7, we referred to small-scale experiments in civic ecology practices, and how learning from the results improves those practices. Now scale this experimentation, measuring, and learning up to multiple practices and organizations. When more than one group is involved in community gardening, each community garden is its own experiment, but it is also part of a larger network. Other organizations in the polycentric governance system, so-called bridging organizations like Urban Harvest or Green Houston, can help assess the costs and benefits of particular strategies, and share what they discover through the network of community gardens and organizations that they support. In this way, the community gardening practice advances, and the polycentric governance system becomes more than the sum of its parts.

Polycentric governance systems enable a diversity of options for civic engagement activities that provide ecosystem services. The ability to draw on diverse organizations and practices may be particularly important when cities and societies more broadly are faced with disturbance and disaster, such as a devastating earthquake, the loss of a key industry, and the droughts, floods, sea level rise, and human migrations brought about by climate change.[7] After Hurricane Sandy, a nonprofit surfers' club whose members frequented New York's Rockaway beaches was among the first groups on the scene, offering food, water, and much-needed supplies before city government was able to reach the stricken beach-front neighborhoods.[8] And hundreds of pocket parks and community gardens offered a place for individuals to express their grief and hope after 9/11; these groups formed a network of Living Memorials supported by the U.S. Forest Service,[9] and provided immediate ways for people to recover while awaiting the formal 9/11 monument that would not be built until a decade later.[10] In cities, civic ecology practices are part of polycentric systems of formal parks, informal green spaces, and stewardship activities. Together, these green spaces and practices provide multiple options for people to connect with nature and with each other, to provision ecosystem services, and to work toward the common good.

Ostrom points out: "No governance system is perfect, but polycentric systems have considerable advantages given their mechanisms for mutual monitoring, learning, and adaptation of better strategies over time. Polycentric systems tend to enhance innovation, learning, adaptation, trustworthiness, levels of cooperation of participants, and the achievement of more effective, equitable, and sustainable outcomes at multiple scales."[11] In short, polycentric governance provides room for organizations working at all levels to have a hand in the game. And just as organizations at the national and global levels contribute in unique ways, local groups contribute in ways that other organizations cannot.[12]

### Antagonism Blues

Polycentric governance means including a rich diversity of organizations and people in making decisions about a particular place or resource. But often the various parties have different and even conflicting priorities about use of that resource. In fact, resource management has a history of battles over shared resources—those favoring resource extraction and development battling those favoring conservation and recreation. In the Pacific Northwest, the schism between loggers depending on forests for their livelihoods and conservationists desperate to preserve remnant old growth trees led to extreme, even terrorist behaviors—like when the Earth Liberation Front set fire to a University of Washington plant-breeding facility allegedly colluding with the forest industry, and when Earth First "tree spikers," hoping to save the spotted owl, pounded nails into old-growth trees to sabotage the work of chainsaws. But after years of conflict the antagonists in the old-growth forest controversy grew weary and looked at ways to circumvent their differences—to come to some sort of truce or even collaboration based on common rather than fractious values.[13]

Nowhere is this story from antagonism to collaboration better illustrated than in the case of the Friends of the Los Angeles River. When *Los Angeles Times* writer and urban explorer Dick Roraback embarked on an expedition to rediscover the Los Angeles River in 1985, he scarcely recognized what he saw as a river.[14] He wrote: "It hasn't any whitecaps. It hasn't any fish. . . . Just to see one ripple would be my fondest wish."

Paraphrasing Woody Guthrie's optimistic depiction of power generated from dams on the Columbia River ("Roll on Columbia, roll on"), Roraback bemoaned how the Los Angeles River "just hauls its load of sad debris from the sewage pipes to the mighty sea. . . . Ooze on, Los Angeles River, ooze on."[15]

By the mid-1980s, industrialization and flood control projects had transformed what once was a free-flowing river bordered by trees—and described by an eighteenth-century explorer as "a very lush and pleasing spot, in every respect"—into fenced-off forbidden territory. The earlier river was now a channel that served two purposes: waste transport and flood prevention. One Army Corps engineer taking credit for successful flood control proclaimed it was "the River we built." The very meaning of the Los Angeles River became so contested that at one point in a public meeting, a verbal battle erupted. A county public works official repeatedly insisted that the river was not a river but a flood control channel, while river activist Lewis MacAdams persisted on calling the river, "*the river*." A state assemblyman countered that the now straightened channel could become a "bargain freeway" for trucks and cars, thus alleviating Los Angeles' notorious traffic congestion by 20 percent.[16]

Although MacAdams originally used 1960s-style protest to advocate for the Los Angeles River, he later turned to nearby Occidental College for help in building a more "friendly" effort to "re-envision the Los Angeles River."[17] Together with artists, local residents, community activists, and urban designers, MacAdams and his allies, now calling themselves Friends of the Los Angeles River, planned a series of events. The events would engage the neighboring communities in reimagining a Los Angeles River that would no longer be a landscape of waste, concrete, danger, and violence, but an environmental and community asset. At the opening session of the re-envisioning initiative, a more environment-friendly government official demonstrated a sense of hope tempered with realism in describing the river's future: "While the River might not be 'the (trumpeter) swan in LA's future, it could be a very, very pretty duck'"[18]

Shortly thereafter, a new vision of a river, which had started out with the provocative actions of a handful of community activists, had the support of mayoral candidates running for office, legislators, and voters. A coalition was established to conduct research and plan for the restoration of the Los Angeles River. Although the struggle between developers and activists continued, significant renewal efforts were implemented, including establishing recreational pathways and access for boating, and restoring the streamside habitat to aid in flood control and provide habitat for wildlife and human enjoyment.[19]

Coalitions of organizations working to integrate civic renewal and environmental restoration, like that established around "reinventing" the Los Angeles River, provide an alternative to the polarized interest-group politics that we saw dividing business and environmentalists thirty years ago, and that still rears its contentious head today in the acrimonious discussions and failure to come to consensus on hydraulic fracking, greenhouse gas emissions, and other fractious issues. These coalitions are part of a collaborative brand of polycentric governance called "civic environmentalism," where civic ecology practices are one player at the table, working alongside groups focused on policy setting, advocacy, and management at larger scales.[20]

Volunteers collect trash along the Los Angeles River, running through a concrete channel and restored streamside habitat. Photo credit: *The Military Engineer*, http://themilitaryengineer.com/tme_online/2012_july_august/la_river/LA-River-4 .jpg. Photo courtesy of USACE

## The Bees and the Trees

Civic environmentalism is not only a type of governance but also part of a larger civic renewal movement—a movement where innovative practices linking environmental restoration with civic revitalization are bubbling from the ground up. These innovations emerge when, after years of being locked in combat, public policy adversaries realize they may be winning (or losing) the battle but definitely are losing the war. And they emerge when it becomes evident that navigating through a labyrinth of government regulations and procedures does not address local issues that people care deeply about. Into the vacuum step the watershed stewards, the urban farmers, the tree planters, the friends of parks organizations— the so-called civic ecology "social entrepreneurs."[21] What do people who study social entrepreneurs and social innovations have to contribute to our understanding of environmental governance?

Social innovations—the new ideas and policies that address social issues such as health, education, and the environment—can be thought of as an "uneasy symbiosis of 'bees' and 'trees:'" "The bees are the small organisations, individuals and groups who have the new ideas, and are mobile, quick and able to cross-pollinate. The trees are the big organisations—governments, companies or big NGOs—which are poor at creativity but generally good at implementation, and which have the resilience, roots and scale to make things happen. Both need each other, and most social change is an alliance of the two."[22]

Civic ecology practices are the "bees"—the small groups that recognize a need in their community, act quickly, and "cross-pollinate" to share novel approaches to rebuilding community and restoring ecosystems. They generally match their creativity and commitment to features of the local place where they live, work, and practice stewardship—including its social-ecological memories, cultural traditions, and a vision for the ecosystem services it could provide—all of which shape the social innovation. Houston's Alabama Street community gardeners, New Orleans activist and tree planter Monique Pilié, the Capuchin monks forging a viable civic agriculture out of vacant land in Detroit, and Kazem Nadjariun who sparked the Nature Cleaners group in Iran are the bees. These are the civic ecology stewards and the social entrepreneurs who ignite the spark for transformations we are seeing across North American and international cities and beyond.

As groups such as Friends of the Los Angeles River and Occidental College join forces with artists, poets, urban planners, engineers, and community activists from Los Angeles's Chinatown to convert wastelands to navigable rivers and riverside parks, they are forging social trust and civic engagement that leads to other types of civic renewal. Perhaps having heard about how greening elevated rail lines in Manhattan, and how daylighting rivers in cities from Seoul, South Korea,[23] to Yonkers, New York,[24] has revitalized downtowns by spawning art shows, farmers markets, recreational walking, nature study classes, museums, and restaurants, MacAdams recently teamed up with architects to plan a project that would convert railroad yards into a park along the Los Angeles River.[25] As these stewardship innovations take root in the city, municipal governments and businesses—the "trees"—come to recognize and support them, and a polycentric governance system better able to respond to future challenges emerges. What is happening in Los Angeles is happening in different forms in cities and watersheds across multiple continents—civic ecology practices partnering with community organizations, environmental nonprofits, and government to build ever-expanding networks, and eventually coalescing into a civic environmental movement.[26]

**Summing Up**

South African resilience scholar Reinette Biggs and her colleagues help us to keep moving forward in stating: "Innovation is always defined relative to a particular context and time. What is new and innovative today is set to become old and the source of new problems in the future."[27] Hence arises the need to not rest on our laurels, but rather to continuously evolve management practices and governance institutions based on what we learn from experience and from more formal research. The abilities to innovate and to use such innovations to adapt and transform become ever more important mandates in the face of accelerating climate, environmental, and societal change.

The flexible creative bees and more rooted trees—the civic ecology stewards and the larger government agencies and nonprofit organizations with whom they partner—are both needed to address civic and environmental decline in Houston and Los Angeles, and in the townships of South Africa and the villages of South Korea. Together they form a polycentric governance system, which enables multiple "experiments" that can be tested, allowing the nonstarters to be weeded out and the shining stars to spread. They suggest that in addressing civic crises that are joined at the hip to environmental degradation, the "store of governance tools and ways to modify and combine them is far greater than often is recognized."[28] How might global and environmental policymakers leverage the full suite of tools—including civic ecology practices—in transforming dystopic communities? We address this question in the final chapter, chapter 10. But first we take another look at systems, this time moving up from polycentric governance to more encompassing social-ecological systems.

# Steward Story: Nam-sun Park

When Nam-sun Park finished elementary school in the early 1970s, her father decided that she should go to work in a factory to help support her family.[1] For Park, poverty not only meant missing out on a secondary education, but also going out into the woods to collect herbs and other plants her family could eat. Today she still remembers what she learned from collecting herbs and shares her knowledge with students and their mothers.

In the mid-1990s, Park had her first taste of environmental activism in her city of Donghae, South Korea. Troubled by garbage in her apartment building, she worked with her neighbors to develop a means to reduce food wastes, and then engaged with a town government official to promote garbage reduction more broadly. Building on this initial success, she founded the local "Environment Love Society," whose membership grew rapidly from three to one hundred members. The Society's activities have evolved over the years, from recycling used clothes into bags and hair accessories, to conducting a Saturday Green School for students and older Donghae residents, and more recently working to protect a local cave.

Having gained local recognition as head of the Environment Love Society, Park was enlisted by the Kangwon Provincial Governor to lead sustainability efforts for United Nations Local Agenda 21.[2] Unfamiliar with what this might entail, Park asked her brother-in-law: "'What is that about?' He answered me: 'It's a selected local agenda which is worked on by citizens, city officers and experts together.' I thought that was exactly what I wanted to do for an ideal democracy in Korea."

As vice president for Local Agenda 21 in Kangwon province, she enacted environmental ordinances and then initiated "Blue Donghae 21" in her hometown Donghae. According to Park, "I know Donghae well since

I am a native here. My experiences from life and the local Kangwon policies helped me a lot. I could see what others don't see as a local agenda."

Park recalls a massive forest fire in 2000, which left nearly twenty-four
thousand acres of forest in ashes and forced coastal residents of Donghae
to evacuate by boat. Two years later, Typhoon Rusa struck Donghae, rain-
ing down 80 cm of precipitation in one night. Dead bodies along with
debris floated onto the beach. A grenade, having floated down the coast
from the North, exploded. The house in front of Park's was "attacked by
a wave and only the old woman who used to be a diver survived. I took
care of her for a month at my home . . . I learned about the fright of natural disasters with my own eyes." Again, her experiences compelled Park to
take action. "On the spot, I directed the distribution of relief supplies to
damaged villages, to volunteer education, and to removing rubbish heaps
on the seashore. Because I played a leading role, people asked me what to
do next. I was not paid for this work, but I did it spontaneously because
it was my village. And I did it from a sense of responsibility that I should
prevent secondary damage."

Despite the massive devastation wrought by the fire and typhoon, municipal officials rebuffed Park's request for funds for planting trees, argu-
ing that there were more urgent needs. But Park was able to secure funds
from the Korean Forest Service's newly established village grove restoration project. According to Park: "For ten years Donghae city had experienced multiple natural disasters such as forest fire, typhoons Rusa and
Maemi, and hail. I had worked connected with those disasters and what I
did for the past ten years was planting trees. For example, when I received
one hundred million Korean won for a Better City Project fund, I spent
seventy million for tree planting. I was that much crazy about trees."

Since reconstruction of the seaside village grove, coastal plants have
grown up and residents are able to see plovers, whereas before only mag-
pies and crows flew noisily overhead. The residents' association for the
city's Bugok district takes care of the trees and conducts annual ceremonies to remember their ancestors at the seaside grove.

Over the past ten years, Park has spearheaded multiple restoration
projects, drawing resources and supplies into Donghae city through gov-
ernment-funded development programs. In addition to the Bugok district
and similar village grove projects in Seungji and at the Yakcheon-temple,

her efforts have sparked trail reconstruction in Chorokbong, as well as the restoration of the Yeondang swamp in the Bupyeng district.

Although her activism over the years has at times sparked controversy, Park remains steadfast in her desire to learn and to move forward. Not only does she remember the lessons she learned from collecting woodland herbs as a young girl, she also is making up for the time when she was denied a secondary education. Park is currently pursuing her doctoral degree. She also reflects on how her local networks and the commitment of village residents have made a difference: "I know the value of community work through the hands of people. My work is not finished yet. These days, I'm thinking about my next step."

# 9

## Resistance, Remembrance, Revolt—and Resilience

*Civic ecology practices are embedded in cycles of chaos and renewal, which in turn are nested in social-ecological systems.*

Award-winning conservationist Theo Manuel was found murdered in his home in Cape Town, the *Cape Times* reported on 22 January. Theo was a valued member of the Fynbos Forum. In 2006 he was awarded a Silver medal by the Cape Action for People and the Environment (CAPE) for his PhD work on the attitudes of the communities of Mitchell's Plain and Khayelitsha to the Wolfgat Nature Reserve. Theo had extensive conservation experience in several government departments, in addition to working as a maths and biology teacher in Mitchell's Plain. He had recently established a private environmental consultancy. Theo was much admired for overcoming major physical disabilities to reach his goals.[1]

Let's face it, the universe is messy. It is nonlinear, turbulent, and chaotic. It is dynamic. It spends its time in transient behavior on its way to somewhere else, not in mathematically neat equilibrium. It self-organizes and evolves. It creates diversity, not uniformity. That's what makes the world interesting, that's what makes it beautiful, and that's what makes it work.

There's something within the human mind that is attracted to straight lines and not curves, to whole numbers and not fractions, to uniformity and not diversity, and to certainties and not mystery. But there is something else within us that has the opposite set of tendencies, since we ourselves evolved out of and are shaped by and structured as complex feedback systems.[2]

Theo Manuel's life in South Africa spanned two worlds—a world marred by extreme violence and dire poverty, and a world marked by the wild beauty of the wind-swept sandy finger of land known as the Cape Flats. Extending north from Cape Town's fabled Table Mountain, the Cape Flats today is home to densely populated and impoverished townships— interspersed with small nature reserves.

As a child growing up in the 1950s, Manuel experienced the forcible eviction of Cape coloreds from Cape Town. His family was resettled in Athlone, one of the many Cape Flats resettlement communities for blacks and coloreds. Athlone later became notorious as the site of the Trojan Horse Massacre. In 1985, South African security forces, concealed in wooden crates at the back of a delivery truck, leapt from their hiding place and gunned down three young men. Thirteen children and adults also were injured in the ambush. The violence was caught on camera and broadcast worldwide, helping to spur the international outcry for an end to apartheid.[3]

The apartheid South Africa in which Manuel grew up was characterized by vicious cycles of racism, poverty, and violence. After the South African apartheid government collapsed under mounting pressure from inside the country and from the international community, the door opened for Nelson Mandela to emerge from twenty-seven years in the notorious Robben Island Prison, and become South Africa's leader in the first election where all South Africans had the right to vote. In the "new" South Africa, numerous barriers for coloreds and blacks to getting an education and breaking out of a cycle of poverty were dismantled. But lingering inequalities have reinforced ongoing vicious cycles of poverty, mistrust, and crime, including the violence that led to Manuel's tragic death in 2008.

Manuel, through his work to foster environmental stewardship in the Cape Flats, contributed to a different kind of cycle. This virtuous cycle includes citizens' groups in the Cape Flats townships who remove trash, help migrating toads avoid cars, clean oiled penguins, cut back alien species and replace them with indigenous trees, and restore the native *fynbos* ecosystem. Through these civic ecology practices, the people of the Cape Flats connect with each other and develop trust. This trust may help to lower crime and strengthen their communities, which in turn could create opportunities for more people to become engaged in stewardship activities. This is what community organizers like Manuel hoped would become the Cape Flats virtuous cycle of greening, reconciliation, and civic renewal.[4]

One of the nature reserves dotting the township settlements of the Cape Flats is Edith Stephens Wetland Park. The park has been described as "a small jewel: a sparkling stretch of water where egrets and cormorants congregate, a wetland teeming with frogs and aquatic life, a centre buzzing with environmental education and other activities."[5] Although only thirty-nine hectares in size and surrounded by townships with dense substandard housing, Edith Stephens is the only place on earth where an endemic species of fern still clings on to life. Members of the surrounding townships have played an important role in shaping the park—through removing invasive species, planting a vegetable garden, and gathering to watch the wetland birds. Edith Stephens Wetland Park is also neutral ground where rival warring gangs from the surrounding communities have come to negotiate and to create relationships across ethnic and racial divides. Reserve manager Luzann Isaacs described the role of the park in reconciliation:

Because Edith Stephens Wetlands Park is in the middle we see it as a meeting ground for all these groups to come together. In the old apartheid system, the railway was designed to keep people separate: coloured people on one side and black people on the other. In the beginning it was very hard to do a project where the two worked together so it has taken us very long but it is happening . . . the youth are starting to speak to each other and create relationships across the railway line and they have adopted Edith Stephens as the place to meet.[6]

The civic ecology stewards working in the harsh social conditions of the Cape Flats townships have faced numerous ups and downs and challenges. Yet Edith Stephens Wetland Park has persisted in its attempts to become a virtuous cycle of greening, reconciliation, and community. What role might Edith Stephens and other small green "jewels" embedded in the Cape Flats townships—where poverty, violence, environmental degradation, and despair have persisted for multiple generations—play in transforming larger social-ecological systems?

## Adaptive Cycle

The five-hundred-square-kilometer Cape Flats,[7] as well as the thirty-nine-hectare Edith Stephens Wetland Park, are social-ecological systems. Throughout this book, we have used the term social-ecological to emphasize the tight connections between the processes inherent to human societies and those that occur in the ecosystems of which we are a part.[8]

An important concept guiding social-ecological systems thinking was developed in the 1970s by ecosystem scientist Buzz Holling, who watched as a devastating infestation of the spruce budworm moth ravaged the forests of Eastern Canada. What had been a mature evergreen forest was transformed by the insect into a landscape with thousands of dead trees. Over time, the snags toppled over to form patches where sunlight was able to reach the forest floor, and as their roots died, competition for soil nutrients eased. These openings in the forest enabled new seedlings to take root. Eventually, young seedlings grew into saplings, which in turn, grew into a forest. Inevitably a new disturbance—be it insects, drought, clear cutting, or a warming climate—once again will decimate the forest and a cycle will begin anew. Holling coined the term "adaptive cycle" to refer to this sequence of rapid growth followed by the more stable or "conservation" phase, and then collapse, release, reorganization, and once again rapid growth.[9]

The "release" phase of the adaptive cycle is a chaotic period following the destruction of the forest by an invasive insect, of an urban social-ecological system such as New Orleans by a major storm, or of a nation such as South Africa by the collapse of apartheid. But chaos opens up windows of opportunity—for social-ecological systems to reorganize.[10] It is a period ripe for novelty and innovation. It also is a period when small experiments—including experiments in community greening and reconciliation such as those in Edith Stephens Wetland Park—emerge. And it is a period during which previously suppressed or less common organisms gain in abundance—such as early successional hardwoods in the spruce-fir forest or trees planted amid the devastation caused by a tornado striking Joplin, Missouri. Eventually, the more promising innovations and newly appearing species take off—resulting in a period of rapid growth and a return once again to the conservation phase.

For a system to maintain itself in the conservation phase, it must have the capacity to adapt. Agricultural systems have adapted to the pressures brought about by a burgeoning population through practices such as intensive tilling and spraying pesticides. While helping the farming system to adapt to changes in the short term, these practices may result in new longer-term stresses—like soil erosion and nutrient depletion.[11] And as systems mature, they can lose options—applying more pesticides no longer controls insects. They also lose adaptive capacity, the ability to adapt to a new disturbance or stress. Some describe these systems as "brittle"— subject to shattering apart when confronted with one more stress.[12]

The adaptive cycle implies that as multiple stresses build up—and when a major disturbance occurs—small adaptations will no longer suffice. A system unable to adapt to ongoing change, and overwhelmed by massive disturbance, reaches a tipping point. Once the system crosses a threshold and enters into the chaotic or release phase, and ongoing adaptation is no longer a viable option, the ability to transform more radically is critical to move into the reorganization phase.[13] Herein lays a paradox: actions taken to ensure stability in the short-term may over the long run hamper the ability of the system to adapt. We revisit the notion of adaptation—and transformation—shortly.

## Trapped

Within an adaptive cycle, one may find multiple smaller processes operating at shorter time scales. These include virtuous and vicious cycles, which can be embedded at different points along the adaptive cycle. After a system collapses, virtuous cycles of greening and renewal may emerge and grow, helping to steer the system toward a period of reorganization. But persistent vicious cycles of poverty, unruliness, and limited economic opportunities—so-called "poverty traps"[14]—also may emerge to fill the void. In these vicious cycles, not only poverty and crime, but also a poverty of ideas, vision, and political will—critical to innovation and transformation—reinforces the trap.[15] Like other systems that have collapsed, the poverty trap system reorganizes but in a manner that is undesirable and forecloses future opportunities for more virtuous cycles. Chaos followed by rebuilding of less desirable forms of organization unable to steer their societies in virtuous directions, appears to be happening in some post–Arab Spring countries today. New vicious cycles are replacing those that preceded them and thus foreclosing options for a more "virtuous cycle" future.

A system that moves beyond chaos and enters into the reorganization, growth, and conservation phases eventually runs the risk of being stuck in a different sort of trap. "Rigidity traps" occur in the conservation phase, when the influx of new species and new ideas is suppressed. Citizens may organize to prevent foreigners from moving into their neighborhoods or governments may impose more and more rules restricting freedoms. These more homogeneous social-ecological systems resist change and lose the capacity to adapt.[16] But eventually another shock or catastrophe opens up a window of opportunity for new ways of thinking. A way out of the rigidity trap emerges—and the possibility of new options for the future.

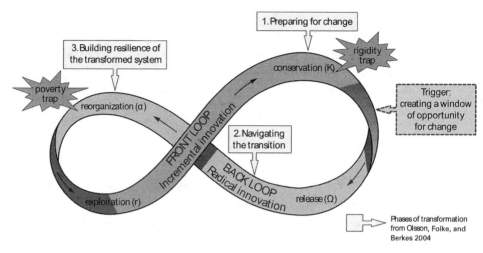

Phases of transformation from Olsson, Folke, and Berkes 2004

The adaptive cycle is a useful metaphor for understanding change over time in social-ecological systems. A "trigger" following the conservation phase creates chaos but opens a window of opportunity for new ideas. Poverty traps occur when an absence of new ideas prevents systems from rebuilding. Rigidity traps result from resistance to innovation by bureaucracies or groups with vested interests in the status quo, and limit a system's ability to adapt to ongoing change. Adapted by Biggs et al. from *Panarchy* by Lance H. Gunderson and C. S. Holling, editors. Copyright © 2002 Island Press. Reproduced by permission of Island Press, Washington, DC, and R Biggs. R. Biggs, F. Westley, and S. Carpenter, "Navigating the Back Loop: Fostering Social Innovation and Transformation in Ecosystem Management," in *Ecology and Society* 15, no. 2 (2010), http://www.ecologyandsociety.org/vol15/iss2/art9/.

## Shocked into New Thinking

After Hurricane Mitch ravaged Honduras in 1998, community leader Doña Justa Nuñez saw an opportunity to address the deforestation that was partly to blame for the landslides that killed her neighbors. Justa Nuñez also connected the deforestation with growing health problems caused by villagers inhaling smoke from cook stoves. Enlisting the aid of NGOs, she organized a women's group in the town of Suyapa to adapt fuel-efficient stoves to the women's cooking traditions. The resulting Justa cook stove is an insulated box constructed from local brick or cinder block and has a built-in chimney that vents smoke from the kitchen. The Justa stove lowers greenhouse gases and pollution inside houses. And because the Justa stove reduces local fuelwood consumption, fewer trees are being felled for fuelwood.[17]

Former U.S. EPA Chief Administrator Lisa Jackson described how another hurricane "shocked" and transformed a woman's environmental perspectives—in ways that helped prepare her and her New Orleans community for a subsequent catastrophe.

Before Katrina, my mother would not have called herself an environmentalist. Despite my chosen profession, she saw environmentalism as the fight for spotted owls, polar ice caps and uninhabited wilderness. She did not see it as part of her life, and she never thought to join the debate on environmental issues.

After Katrina, my mother saw the reports that the flooding was exceptionally destructive because our marshes and wetlands—the area's natural and most effective defenses—had been destabilized by navigation channels, covered over by levee construction, and most damagingly, cut away for the placement of oil and gas lines. . . .

Five years later, this lesson is taking hold. My mother has joined other unlikely advocates in the call for wetlands restoration and protection. Thanks to environmental heroes like Brad Pitt and Wendell Pierce and organizations like Global Green, rebuilding projects in New Orleans are focused on sustainable design and clean energy—and the green jobs that come with them. My old neighborhood, Pontchartrain Park, is being redeveloped and reborn as a sustainable community and model of new green urbanism, including my mother's house. . . .

With the lessons of Hurricane Katrina in mind, we immediately involved environmental justice and community advocates when responding to the Deepwater BP oil spill.[18]

Hurricanes in Honduras and New Orleans resulted in massive destruction and chaos, but also created opportunities for innovative thinking to emerge. New understanding of how tightly linked humanity's future is to that of nonhuman nature helped to build a stewardship ethos, and to propel changes in fuelwood consumption in Honduran villages and restoration efforts along the Gulf Coast.

## Adaptation and Transformation

More frequent than large-scale catastrophic disturbances like hurricanes are smaller disruptions like cutting trees for firewood and paving over a vacant lot. Civic ecology practices can be part of ongoing adaptation to smaller changes. Volunteers installing bioswale gardens in Seattle are helping to reduce runoff into Puget Sound. In this way, they are part of the city's effort to adapt to increased area in paved surfaces that are impermeable to rain. Such small-scale adaptations are common when changes accumulate gradually and do not involve a major disturbance.

Civic ecology practices also kick in after systems experience drastic change—sometimes with a new urgency and intensity.[19] After the fall of communism in East Germany, citizens and government worked together to convert the fortified strip of land occupied by the former Berlin Wall into a greenway.[20] And following the devastation wrought by the atomic bomb in Nagasaki and Hiroshima, people in Japan came together to preserve the few surviving trees.[21] Similarly, after the war among the Balkan states that comprised the former Yugoslavia, community gardens emerged that brought together people from opposing sides of the conflict.[22]

In short, civic ecology practices play a role not only in adaptation to ongoing change, but also in the transformation that is needed after a major disturbance tips the system into the release or chaotic phase of the adaptive cycle. Sometimes the same practice contributes to adaptation and transformation at different times. Historically, villagers in South Korea planted small groves of trees to help protect against wind and erosion; after a devastating typhoon, villagers set about to replant the village groves that had been destroyed, receiving support from government and an NGO working on similar restoration projects in villages throughout Korea.[23] In chapter 10, we ask: How can scientists, NGOs, and government help small innovations such as village grove restoration "go viral," and spread out through a city, region, or country?

## Resilience Thinking

Social-ecological systems resilience refers to the capacity of a social-ecological system to continually change—or adapt—so as to maintain ongoing processes in response to gradual and small-scale change. It also captures the ability of a system to renew and reorganize—to transform—when faced with devastating change.[24] Thus, the notion of social-ecological systems resilience encompasses two separate but related processes encapsulated in the adaptive cycle—adaptation in the face of ongoing change and transformation once a system has crossed a tipping point or threshold and fallen into a chaotic state. Change or disturbance may be in the form of major storms, fire, pollution, global warming, economic decline, or war and political upheaval. Whether or not such disturbances represent devastating disasters depends in part on the system's adaptive capacity. For example, a city harboring a diversity of cultures, ideas, and livelihoods would be expected to have a greater capacity to adapt to the shutting down of a business than a city dependent on one industry.

A seminal moment in resilience thinking was a paper presented at the 2002 World Summit for Sustainable Development in Johannesburg, South Africa. There Swedish resilience scholar Carl Folke along with twenty-four other scientists described how a monoculture, such as a plantation of fast-growing poplar trees used for biofuels, lacks diversity and thus may be unsustainable in the face of disturbances such as insects and plant diseases. A further problem is that such monocultures fail to sustain communities that depend not just on fuel, but also on multiple ecosystem services provided by mixed-species forests, such as food, freshwater, flood regulation, and opportunities for cultural expression. In contrast, systems that harbor a diversity of species and incorporate diverse stakeholder perspectives into decision making are likely to be more resilient in the face of both small-scale change and catastrophic disasters. Because change occurs in all systems, resilience is an integral component of sustainability.[25] The relationship of resilience to sustainability is further spelled out by Folke and his colleagues:

Managing complex, coevolving social-ecological systems for sustainability requires the ability to cope with, adapt to and shape change without losing options for future development. It requires resilience—the capacity to buffer perturbations, self-organize, learn and adapt. When massive transformation occurs, resilient systems contain the experience and the diversity of options needed for renewal and development. Sustainable systems need to be resilient.[26]

In short, managing for ongoing and inevitable change, rather than toward a stable or "sustainable" endpoint, is at the core of resilience thinking.

The resilience framework has roots in systems dynamics, which emphasizes feedback loops such as vicious and virtuous cycles, in which each event "feeds back" to reinforce the previous one. This is in contrast to more linear thinking, which focuses on cause and effect. Resilience scholars would ask: What are the multiple interactions of poverty, crime, low levels of education, and environmental degradation? In contrast linear thinkers might inquire: Does low educational achievement cause poverty? According to systems scholar George Richardson, "Systems dynamics thinking gets a lot of its power from a 'feedback' perspective—the realization that tough dynamic problems arise in situations with lots of pressures and perceptions that interact to form loops of circular causality, rather than simple one-way causal chains."[27]

But what differentiates a resilient system? Attributes that enable systems to adapt to ongoing change and thus avoid crossing thresholds are similar to those that allow systems to embark on a desirable path of reorganization or transformation following a catastrophic, tipping-point disturbance. [28] Many resilient system attributes also are similar to the pieces that hold together civic ecology practices, including social capital, biodiversity and ecosystem services, and polycentric governance. Other attributes include cultural diversity (bringing multiple perspectives to the table helps us find pathways out of quandaries[29]) and innovation and experimentation. [30]

Resilient systems also allow self-organization, or larger processes to emerge from smaller ones in a seemingly unplanned manner. Nutrient cycling in ecosystems is a process that emerges from the actions of millions of microbes, insects, and larger organisms, each going about consuming and breaking down organic wastes.[31] Civic ecology practices also are self-organized when, without direction from higher levels of governance, individuals act together to convert a vacant lot to a community garden, daylight a paved-over stream, or transform concrete basins into ponds for dragonflies and fish.

Finally, in resilient systems, people learn through observing and reflecting on the ways in which they manage their resources, and then applying what they learn to improving their management practices.[32] When a system has crossed the threshold into the chaotic phase, what is needed is a more advanced form of learning, one that questions commonly held assumptions. A community that has seen the fisheries upon which their livelihoods depend collapse several times may question whether it's wise to simply switch fishing tactics once again. They may reflect more deeply on their future options—perhaps imagining alternative scenarios such as setting aside marine sanctuaries that will provide future "biological memories" for their fisheries, or foregoing fisheries for ecotourism. Environmental scientist Donella Meadows and her colleague Peter Marshall reinforce this point: "Mental flexibility—the willingness to redraw boundaries, to notice that a system has shifted into a new mode, to see how to redesign structure—is a necessity when you live in a world of flexible systems."[33]

Whereas Holling and his followers talk about resilience of social-ecological systems, the notion of resilience also helps us to understand how people who face overwhelming adversity sometimes exhibit not only the capacity to maintain stability in their lives, but also to grow from devastating experiences.[34] In talking to survivors of violent and chaotic "red zones"—whether they be war refugees taking up a new life in Toronto or survivors of Hurricane Katrina in New Orleans's 9th Ward—we often hear stories about how the act of cultivating—flowers, vegetables, trees, or habitats for fish, wildlife, and birds—has been critical to emotional survival and to engendering hope for the future. Gardening and other forms of stewardship help soldiers and civilians survive war and reintegrate into everyday society, similarly demonstrating how people turn to nature as a resilience strategy in times of stress.[35] Thus, parallel to social-ecological systems resilience, human resilience captures notions of recovery and rebuilding following crisis and hardship. This leads us to the notion of nested systems of resilience, from the individual to the community and larger social-ecological system.

## Nested Systems

In his book *Nested Ecologies*, Edward Wimberley describes his vision of living beings, ecosystems, and the cosmos as a series of ecologies, the smaller ones nested within the larger like a progression of Russian dolls. He thus provides a framework to understand the multiple levels at which civic ecology practices operate.[36]

According to Wimberley, individuals like Manuel in the Cape Flats have a personal ecology—their interactions with people, nonliving things, and other life surrounding them. Nature stewardship can be part of one's personal ecology and a means of individual resilience—of restoring emotional and physical well-being during chaotic and violent times.

An individual's personal ecology is nested in the next larger Russian doll—a community ecology of interactions among family and neighbors. Individuals who join together to restore and steward their environment develop trust and form new ties, thus creating social capital and a sense of community. In the Cape Flats, these community outcomes extended beyond the volunteers at Edith Stephens Wetland Park, as local youth came to the preserve to help resolve their differences.

The next largest nesting doll embraces the personal and community "ecologies," and expands to incorporate interactions among the animals, plants, and wetlands that constitute an "ecosystem ecology." The volunteers at Edith Stephens Wetland Park cultivated vegetables in community gardens and helped to preserve an endangered species of fern.

Wimberley goes on to describe a cosmic ecology, which is perhaps his most important contribution, in that he insists on confronting us with the harsh reality that, just as humans are inextricably embedded in ecosystems, Earth remains inextricably embedded within cosmic systems (our solar system, the galaxy, and ever-larger systems). And just as ecosystems sustain humans, the cosmic system—in particular the sun—sustains life on Earth and largely dictates the conditions under which the planet's life, including humans, exist. Thus, Wimberley's cosmic ecology serves to remind us that despite humans' at times herculean efforts to transcend our biology—to pretend that we are not dependent on or even related to other life—we are completely and utterly dependent on that life, and on our sun. Just as the Earth's biosphere depends on the energy provided by the sun, so are humans dependent on the ecosystem services provided by the biosphere.

Whereas the Russian doll metaphor is helpful in envisioning outcomes of civic ecology practices at the individual, community, and ecosystem levels, we need to think of dolls that are capable of changing form to understand the set of interactions that cut across levels. Here again we draw on the notion of ecology, with its recognition of the interconnectedness of organisms and their physical environment. In an ecosystem that becomes healthier through stewards re-establishing native plants and cultivating gardens, people are more likely to walk, to rest, to reflect, and to enjoy nature. Thus, not only do people's actions enhance communities and ecosystems, healthier ecosystems can change communities and people. The larger Russian doll, through her interactions with the smaller ones, herself experiences changes and helps to change the others.

Related to civic ecology, the important lesson is that individual, community, and ecosystem outcomes are all possible, through stewardship actions starting with individuals and cascading up through higher-level community and ecosystem ecologies. Further changes in ecologies at higher levels—like the Cape Flats ecosystem—can in turn cascade down and change the lower-level community and personal ecologies.

## Panarchy: the God of Creation and Destruction

Thus far, we have talked about vicious and virtuous cycles that depict how actions reinforce each other in seemingly unending loops, and Holling's adaptive cycles that help us understand how social-ecological systems change and respond over time. And we have seen how ecologies—or interactions—operate at different nested levels, from the individual human to the community and larger ecosystem. Through this progression, we encounter increasing complexity—we stumble across words like chaotic, dynamic, feedback, and nonlinear. Here, the idea of hierarchical nesting of different levels of interactions gains importance, but the metaphor of static wooden Russian dolls each encompassing the next smaller but failing to interact, is less useful.

Holling and his colleagues set out to find a term that could convey the complexity of interactions across adaptive cycles operating at larger and smaller scales. They rejected the term "hierarchy" in favor of a Greek god:

> Since the word *hierarchy* is so burdened by the rigid, top-down nature of its common meaning, we prefer to invent another term that captures the adaptive and evolutionary nature of adaptive cycles that are nested within the other across space and time scales. We call them *panarchies*, drawing on the image of the Greek god Pan—the universal god of nature. This "hoofed, horned, hairy and horny deity" represents the all-pervasive, spiritual power of nature. In addition to a creative role, Pan could have a destabilizing, creatively destructive role that is reflected in the word *panic*, derived from one facet of his paradoxical personality. . . . He therefore represents the inherent features of the synthesis that has emerged in this quest for a theory of change.[37]

Panarchy captures the unpredictability and creative potential of complex systems. These are systems where "multiple adaptive cycles occur at varying temporal and spatial scales, nested within and interacting with one another."[38] As opposed to the analogy of nested dolls, each one self-contained in a hierarchical fashion, the image of a Greek god of nature at once creative and destructive—that is, panarchy—attempts to capture the interactions between smaller and larger adaptive cycles. Applying the notion of panarchy to the Cape Flats, one can imagine that the peace talks at Edith Stephens Wetland Park would be noticed by other park managers who similarly are dealing with chaos and violence in the neighborhoods surrounding their park. Seeing the results at Edith Stevens, they might encourage reconciliation efforts at their own parks. As the idea takes off, the

regional governing bodies eventually take note, and incorporate parks—
and community efforts to restore parks—into regional conflict mediation
strategies. Thus, what started in one park may cascade up to a system
of multiple parks and more broadly impact regional governance. In a
panarchy, this phenomenon of small-scale processes cascading up levels
of adaptive cycles to influence larger and slower processes is referred to
as "revolt."

Processes can also cascade down from higher to lower levels in a panarchy. These processes depend on accumulated resources and experience
in higher levels of the panarchy, and thus are referred to as "remember."
Suppose disaffected youth in neighborhoods surrounding Edith Stephens
Wetland Park set off a fire, destroying wetland plants. The ability of the
park to reestablish the plants and continue reconciliation talks is constrained by resources available at higher levels of the panarchy. Has the
larger system of the Cape Flats parks accumulated seeds—the biological
memories—that can be used to reestablish the vegetation that was burned
in the Edith Stephens Wetland Park? Does regional government have ne-
gotiators who retain social memories of how to assist in reconciliation
among warring parties? Together the seeds and reconciliation expertise
would constitute the social-ecological memories that exist across a system of multiple parks at higher levels in the panarchy. These memories
cascade down to the lower levels of Edith Stephens and other individual
parks, thus enabling reorganization or rebuilding after the arsonists' fire.

Because a panarchy has processes operating at different levels, it can
create and test new experiments or solutions, while also preserving and
accumulating social-ecological memories. Again Holling and his colleagues help us understand how this occurs:

> In a healthy society, each level is allowed to operate at its own pace, protected
> from above by slower, larger levels but invigorated from below by faster, smaller
> cycles of innovation. . . .
> The (smaller) fast levels invent, experiment, and test; the (larger) slower levels
> stabilize and conserve accumulated memory of past successful, surviving experi-
> ments. The whole panarchy is both creative and conserving. The interactions be-
> tween cycles in a panarchy combine learning with continuity.[39]

Ideally, policymakers at upper levels would learn from the experiences
of practitioners at lower levels of the panarchy. However, even in relatively
peaceful times, information that flows up from lower levels may be ignored
at upper levels where policies are formulated,[40] as when policymakers

institute incentives for commercial development that destroy civic ecology practices serving as a source of resilience. Thus, positive panarchical change cascading up the levels of nested adaptive cycles "can occur only when a triggering event unlocks the social and political gridlock of larger levels in the panarchy."[41] Most of the chapters in this book explore the role of greening in adaptation and transformation in adaptive cycles at lower levels (neighborhoods, communities) in the panarchy, and thus ignore triggering events that might dislodge political systems at upper levels. In our work on "greening in the red zone," we have described how larger collapses or changes—such as the dissolution of the apartheid system in South Africa and communism in Eastern Europe—unlock larger levels and create opportunities for creativity, reorganization and renewal below.[42] As we move toward a future where climate change makes disturbances and even collapse at all levels of the panarchy more frequent, the question arises: Will the shocks unlock thinking and enable civic ecology practices—the experiments and innovations at the lower levels of the panarchy—to cascade up and help to transform the higher levels?

### Revisiting Athlone

Resilience, in broad terms, refers to the ability of humans, communities, and larger social-ecological systems to rebound and to rebuild in the face of outside stresses, including the death of loved ones, conflict, war, and disaster. During such times of crisis, breakdown, and reorganization, existing and new sources of resilience come to the fore; for this reason, discovering, building, and safeguarding those sources of resilience is critical to future capacity to recover from crisis.[43] Civic ecology practices, as an expression of human will and collective action applied to environmental stewardship, represent a critical source of resilience at multiple levels.[44]

A growing body of research demonstrates the role of greening in individuals' ability to adapt and transform when confronted with hardship.[45] Manuel endured the hardships of growing up amid the violence and oppression of the apartheid era. At age twelve, he suffered a spinal cord injury that hampered his mobility for the rest of his life. Perhaps his interest in green spaces, and in restoring nature and community, were in part a strategy to adapt and grow when faced with these hardships. The subsequent collapse of the apartheid system was a triggering force that

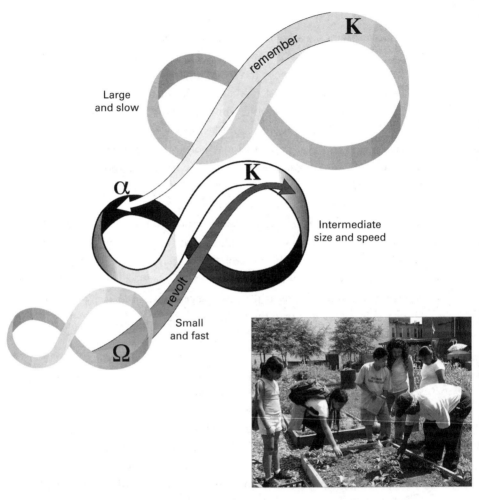

Panarchy provides a metaphor for how adaptive cycles at different levels can be nested within one another. Lower-level adaptive cycles are constrained by processes at higher levels (remember) but may also spiral up to change higher-level processes (revolt). A community garden encompasses adaptive cycles at the individual, community, and ecosystem level. L. H. Gunderson and C. S. Holling, eds., *Panarchy: Understanding Transformations in Human and Natural Systems* (Washington, DC: Island Press, 2002). Image reproduced with permission from Island Press. Photo credit: Alex Russ, Civic Ecology Lab, Cornell University

cascaded down to the Colored Cape Flats communities and residents, and enabled Manuel to get an education and find meaningful work. Tragically, the greening efforts he supported through his research and community work have failed to spread throughout South Africa, and to cascade up through higher levels of organization. Today, vicious cycles of violence, poverty, and environmental degradation persist in the Cape Flats, Soweto, and other South African townships.

Yet given the nascent virtuous cycle of greening and peace negotiations that transpired at Edith Stephens Wetland Park, perhaps it is time for those at higher levels in the panarchy—the policymakers, or the more forward-thinking policy entrepreneurs—to take a second look at civic ecology practices. Could a policy entrepreneur at higher levels institute policies that support these efforts, allowing them to grow and engulf the vicious, poverty trap cycles, and to "revolt" back up the panarchy from individuals to communities to a whole country or nation-state? We address this question in the final chapter.

# Steward Story: Mandla Mentoor

As a child growing up in one of thousands of drab matchbox houses packed together in the sprawling Soweto township, Mandla Mentoor experienced first-hand the insidious waves of apartheid permeating all aspects of black people's lives. When he was just seven, his mother took a job as a housekeeper in an affluent Johannesburg suburb, returning to see her son only every second Thursday. Left to fend for himself, Mandla became unruly—picking fights with neighborhood kids and tossing rocks through their windows. His parents approached a local reformatory to take him in, but in the end sent him to live with an older brother in the countryside. There Mandla joined a Boy Scout troop. He led camping trips and hiked and swam among the craggy peaks and clear streams of the Maluti Mountains. There he also realized his potential as a community leader, and what is possible to achieve when "nature and humanity intersect."[1]

Equipped with these realizations, Mandla returned as a teenager to Soweto—a Soweto in the violent convulsions of a dying apartheid system. Protest theater, underground politics, and stone throwing were all part of the struggle to bring down the oppressive system. But Mandla also connected the stench of the trash heaps permeating the townships with the struggle—and with rebuilding once the apartheid system collapsed.

Mandla Mentoor's vision was that trash could be transformed into something useful—even beautiful. In 1993, just three years after Nelson Mandela was released from Robben Island Prison, Mentoor began "recruiting children and the unemployed to scour the neighborhood collecting bottles, cans, paper, and other bits of waste to make art they could sell."[2] These initial efforts grew into something larger. Mr. Mentoor and his neighbors eventually convinced the local authorities to allow them to occupy a hill looming above Soweto that for years had become a symbol of all that was wrong with South Africa—a dangerous violent place of tire burnings, robberies, and rape, a dumping ground, a no man's land.

Today, from the water tower atop the Soweto Mountain of Hope, one gazes out at the endless conglomeration of matchbox houses like the one Mentoor grew up in. But if one looks downward at the slopes of the mountain, the view is radically different from what a township resident risking her life to cross the mountain would have experienced before 1993. Perhaps the word painted on the water tower—"ubuntu," meaning "togetherness"—best reflects the activity on the slopes of the mountain and at Mentoor's fuschia-and-turquoise-painted house nearby. Together, township residents tend vegetables, medicinal herbs, and fruit and indigenous trees. They gather the hill's rocks to bead them into circles that define the boundaries of spaces used for community discussions and dance performances. And as hearses drive by on their way to the local cemetery, residents cultivate garden plots in the shape of AIDS ribbons to commemorate loved ones being carried to their graves.

Mentoor's vision is larger than one hill symbolizing togetherness towering above Soweto. Spurred by a visit from UN Secretary General Kofi Annan, and famed primatologist and UN Ambassador of Peace Jane Goodall during the Johannesburg World Summit in 2002, and supported by a social entrepreneur fellowship from the prestigious Ashoka Foundation[3], as well as by banks and NGOs, he has expanded the activities at the Soweto Mountain of Hope. Children congregate to learn about water conservation, and young adults prepare for a city council meeting. A recycling center that converts waste into economic assets, an Internet café, concerts, food catering, bike tours for tourists, and a sewing cooperative are all part of the Mountain of Hope today. And Mr. Mentoor has begun to expand youth-driven environmental programs to other Johannesburg area townships. Due to Mandla Mentoor's vision and the efforts of thousands of township youth and adults, the Soweto Mountain of Hope has become an international symbol of greening defiance, resilience, and regrowth.[4]

# Third Interlude

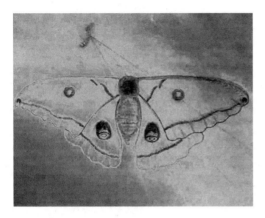

Like butterflies, moths undergo a profound transformation from caterpillar to winged adult. But moths, being night creatures, also symbolize an inner knowing of how to navigate, even in the darkest hours of our lives. The Polyphemus moth pictured here takes its name from one of the giant Cyclopes of Greek mythology, who lived in a fertile land but were blinded to the virtues of their land and to agriculture, and later, as a result of Odysseus's trickery, were further blinded by greed and rage. Our image depicts the possibility of a second, deeper transformation—from a narrow point of view or even blindness, back to a more expansive one, following the rays in the opposite direction to the source of life, the Sun. Art: Keith G. Tidball

Thus far, we have seen how civic ecology practices emerge, we have examined their pieces, and we have learned how they are nested in systems of polycentric governance as well as larger, resilient social-ecological systems. In the final chapter, we explore the last civic ecology "policy" principle. We observe how scientists with a social conscious form partnerships with civic ecology practices, and in so doing, help make the practices more effective. We also come to understand how innovative policymakers—so-called policy entrepreneurs—support civic ecology practices as part of larger efforts to promote sustainable and resilient cities and regions. Critical to the participation of scientists and policy entrepreneurs is their willingness to listen and to be sensitive to the self-organized nature of civic ecology practices—to the community initiative demonstrated by civic ecology stewards. How do scientists and policy entrepreneurs go about helping to grow civic ecology practices?

# IV

## Policy Entrepreneurs: Understanding and Enabling

# 10

## Policy Frameworks: Scaling Civic Ecology Practices Up and Out

*Policymakers have a role to play in growing civic ecology practices.*
Crowning a hill of embassies and apartments in the Sheridan-Kalorama neighborhood of Washington, DC, is a small city park. A mail bin affixed to a post holds copies of the Friends of Mitchell Park newsletter. "Fall Fun Day 2012" is the lead article, showcasing a recent neighborhood event organized by the friends association that featured popcorn, hotdogs, and candy, along with music, a bake sale, flea market, and a table where other community organizations displayed information and answered questions. The fire department brought out a fire truck for kids to climb on, and "an acrobat performed death-defying moves using only a few chairs and a mat."

But turning the newsletter page, one discovers that Friends of Mitchell Park (FOMP) does more than draw children and families to the park on "fun days." An article from the Friends Maintenance Committee describes summer 2012 in the park:

What a summer to have had glorious irrigation! The intense prolonged heat was hard on plant material everywhere in the city, but the Park survived very well. The only notable heat-related casualty was the demise of two linden trees which had struggled since their planting by Casey Trees in October of 2009. Casey Trees will return to the Park on October 23rd to replace these two trees. We are grateful to Casey Trees not only for the earlier group of 17 trees, but for all the plantings and care they do all over the city. . . .

Adding to the beautification is the help of Claire Wagner & Paula Holland's weeding group. These neighborhood volunteers meet every other Sunday to hand weed the flowerbeds. If interested, . . .

The herb garden has been a wonderful success, even attracting new park users who stopped by to collect fresh herbs. We will need to make adjustments as to placement and soil, but we will add new herbs next spring. People have been so thoughtful not to take large quantities at one time, and dogs have not been a problem.

The one problem we do have is a rat population that the city is aware of and has begun addressing through a preventative program, which includes picking up the trash twice a week.

We alerted the Department of Parks and Recreation (DPR) that they need to patch the holes in the playground surface and install an extra picnic bench to support the growing lunchtime population.

The exterior of the recreation building needs $40,000 in repairs which are not yet in the DPR budget. The building is used and staffed in winter and summer by DPR for a morning preschool program. The scope of repair work includes siding, windows and sills. FOMP will continue to lobby DPR for a budget allocation.[1]

The activities of Friends of Mitchell Park illustrate multiple components of a civic ecology approach. Volunteers living nearby have organized to monitor tree health, pull weeds and plant trees, as well as keep an eye on use of the herb garden. Their stewardship and related activities are like a magnet—they draw other volunteers and families into the garden, creating opportunities for community involvement and connecting with neighbors, while also providing cultural (recreation), provisioning (herbs), and regulating (trees providing shade and absorbing $CO_2$) services. Friends of Mitchell Park advocates for park improvements. And they are part of a broader park governance system consisting of their own organization, the City Department of Parks and Recreation, and a private tree care company. Despite all these activities and connections, the impacts of Friends of Mitchell Park are local and limited—realized only on one small patch of land within one city neighborhood. How might the policymaking organizations—government and larger NGOs—scale out the impacts of local civic ecology practices?

Cash-strapped city governments increasingly recognize the value of community organizations and nonprofits in creating green infrastructure and providing ecosystem services. They may see friends of parks organizations and other civic ecology groups as an inexpensive way to get things done—like tree planting or biofiltration projects.[2] And "outsourcing" to civil society organizations can include not only ecosystem services provision, but also other functions that traditionally have been the responsibility of city government. In 1980, civic and philanthropic leaders who were determined to reverse the neglect of "America's first and foremost major urban public space" and restore it to "its former splendor" founded the nonprofit Central Park Conservancy.[3] Eighteen years later, the Conservancy signed a management agreement with the City of New York, assuming responsibility for park maintenance. Now the official management body for Central Park, the Conservancy employs the park's maintenance staff, and raises the lion's share of its $45.8 million annual budget.[4] The Conservancy also provides hands-on opportunities for volunteers to join in their stewardship and restoration mission, through activities like mulching, planting, raking leaves, and sweeping paths.[5]

**Small Beginnings, Broader Impacts**

Even large well-funded friends of parks groups like Central Park Conservancy, or grand restoration projects like the High Line, have small beginnings. A group of citizens and community leaders become alarmed at government neglect of parks and open space that have become eyesores— and broken places. They seize the initiative to restore nature and their community. That these small, self-organized efforts grow to the point where they are seen as important partners for city and state government in providing ecosystem services, and even park maintenance, is a sign of their success. But from a civic ecology perspective, a balance needs to be struck—a balance between the passion and commitment of ordinary people who steward nature and community in their own neighborhood, and the scaling up of such efforts to have impacts at a city or even regional scale. Thus, policymakers who see value in and seek to expand civic ecology practices need to be mindful of ways in which these practices tie together their communities, as well as the local initiative and empowerment that set them apart from other environmental practices.[6] This means that top-down policy directives will not work.

Civic ecology stewards are the social-ecological entrepreneurs—the inventors or the "bees" working at the local level. We can distinguish them from policy entrepreneurs—the "trees" or the people working for government and NGOs who shape the social, institutional, and policy environment to favor the growth and spread of local innovations.[7] Policy entrepreneurs leverage their communication skills, understanding of the larger political and institutional context, and position in social networks to create the environments that enable civic ecology and other innovations to take flight.[8] Policy entrepreneurs concerned about managing for sustainability or resilience see the potential for civic ecology practices to become part of larger, citywide, regional and even global natural resources management strategies. Their challenge lies in first locating and then supporting and expanding these small, self-organized efforts.[9]

Three general steps help move civic ecology practices from local stewardship to broader policy innovations. We begin with a step we have already embarked on—giving a label to the phenomenon (in our case "civic ecology"). We next examine how civic ecology practices can become more effective as local providers of ecosystem services and contributors to community well-being through partnerships with scientists. Finally, we explore how civic ecology practices can expand and spread—as government and NGO policy entrepreneurs shape the environment and serve as a bridge between civic ecology practices and other stewardship efforts to form regional adaptive and collaborative resource management systems.

## The Gallery: Finding a Label for a Collection of Images

Imagine a city councilman wandering through a gallery of photos, each depicting a different civic ecology practice.

- A community garden in the Bronx where Mexican immigrants cultivate *alache, epazote,* and *papalo.*[10]
- Members of a yacht club on Staten Island installing oyster reefs and monitoring the young oysters' growth and survival.[11]
- Young adults from conflicting ethnic groups coming together to plant a reconciliation garden in a South African township.[12]
- Combat veterans working with youth to restore coral reefs.[13]
- Elders in South Korea restoring traditional village groves decimated by a typhoon.[14]
- Sikhs and Mennonites joining together to plant a garden as part of the City of Winnipeg's Adopt-a-Park program.[15]
- Japanese villagers recreating satoyama systems and their urban counterparts reconstructing ponds that attract dragonflies.[16]
- Volunteers coming out for the annual Friends of the Los Angeles River trash pickup and monitoring day.[17]
- Elders from a senior center helping revitalize a neglected city park while their grandchildren clear vegetation to create a greenway linking open space in Chester, Pennsylvania.[18]

• Volunteers with Groundwork-Hudson Valley's Free-A-Tree project removing invasive vines that are strangling trees and shrubbery along the Saw Mill River in New York State.[19]

• And residents of New Orleans' 9th Ward replanting and caring for damaged live oak trees after Hurricane Katrina.[20]

These are just a few of the images that the councilman encounters in the gallery. Maybe the images intrigue him. Perhaps the photos bring to life nascent ideas about the potential for similar efforts in his city. Maybe he is already thinking about ways to improve environmental quality—and quality of life—in his district. And at the back of his mind, he may be worried about how his city might respond to the next shock—the next disaster—be it a heat wave, a flood, or a drought brought about by climate change, or economic upheaval following a major employer shutting down business.

But making connections among all the images is tricky. Why are they all part of the same photo exhibition? What do they have in common?

The images all depict practices in which local people have taken the initiative to steward local resources, most often after disasters or in neighborhoods experiencing social, economic, and environmental stress. However, even if the councilman starts to see connections he may not have a label for the collection of photos. How can he communicate with his colleagues—and with the public who supports his work—about things as diverse as community and reconciliation gardening, oyster restoration, reconstructing ponds to attract dragonflies, and riverside trash cleanups?

Assuming the councilman is a policy entrepreneur who wants to create change through leveraging local solutions to local problems, he may grasp for a label for the diverse set of practices in the gallery. He may realize that similar practices already are taking root in his own city.

Researchers who study how policy entrepreneurs facilitate the spread of social innovations have found that a first step is to create a label, which while intentionally vague, serves to mobilize a political coalition.[21] Examples include "pro-life," "pro-choice," and "feminist." Civic agriculture is a label for practices linking food production to civic renewal, including farmers markets, organic farming, and community-supported agriculture.[22] Similarly, ecosystem services was intended as a label that would change the public's attitudes toward conservation by alluding to, and even placing a dollar value on, the services biodiversity and ecosystems provide for humans.[23]

Sometimes labels can be powerful tools in environmental advocacy, through alluding to an image that frames the way people view a resource. During the late 1980s, when community activist Lewis MacAdams first began advocating for restoration of the Los Angeles River, he persisted in using the term "river" as a symbol for what the concrete-sided channel could become. Public officials referred to the same resource as a "flood control channel" or "bargain freeway." In the end, MacAdams's vision of a river appealed to legislators and the wider public, and likely contributed to the city's rejection of the idea of turning the channel into yet another highway, as well as to the eventual successes of the restoration efforts.[24] Similarly, in Sweden, an environmental activist reframed perceptions of a wetland, from "water sick" to "water rich,"[25] which provided the impetus for local officials to implement environment-friendly policies.

A social or political upheaval provides a window of opportunity for reframing ideas and subsequently revising policy.[26] In the case of the Swedish wetlands, activists took advantage of a window of opportunity provided by municipal officials' concern that they needed a new, more positive identity, and provided them with wetlands conservation as a means to achieve their goal.[27] Similarly, in coastal Louisiana, Hurricane Katrina provided an opportunity to reframe the coastal ecosystems from swamps to valued wetlands.[28] And just after Hurricane Sandy ravaged low-lying areas of New York and New Jersey, the media and environmentalists leveraged the disaster to reframe oyster reefs and coastal dunes from something that could be ignored and flattened to valued resources for protecting communities from flooding and devastation.[29]

And what about our city councilman musing at the images of community gardening, coral reef restoration, tree planting, and other civic ecology practices in the gallery? Or the scientists, foundation directors, and community tree stewards who join him at the exhibition? Will they develop a common understanding—and a label—for all the practices depicted in the photos?

In 2007, we first proposed the term "civic ecology" as a label for self-organized stewardship actions with both community and environmental outcomes, and for the study of the interactions of these practices with other processes taking place in the larger social-ecological system.[30] We specifically chose the term "ecology" to suggest feedbacks and other interactions, and to focus on systems thinking.

Perhaps a label—civic ecology—will help the visitors to the gallery understand the commonalities among the assemblage of practices and their potential impacts. Perhaps our label also will help the councilman, as well as the scientists, NGO staff, foundation director, and community volunteer stewards in the photo gallery, to reflect and even develop a larger vision for actions they can take to foster self-organized practices that contribute to ecosystem services, quality of life, adaptation and transformation, and more broadly sustainability and resilience in their city and beyond.

## Strengthening Civic Ecology Practices

Scientists who have entered the photo gallery may adopt a skeptical stance regarding any undocumented claims about the outcomes of civic ecology practices. How do we measure the impact of a friends of parks association whose members remove invasive species on the reestablishment of native plants and wildlife? How do we know that bioswale gardens are actually retaining water running off streets and roofs, let alone improving water quality in a nearby stream? How can we be sure that the Japanese satoyama system is effective in bringing back biodiversity? And what evidence do we have that gardeners from different ethnicities at Soweto's Mountain of Hope are in fact establishing stronger connections with each other and with nature? Skeptical scientists have a critically important role to play in strengthening civic ecology practices—through research conducted in partnership with civic ecology stewards to measure and enhance their stewardship outcomes.[31] Such scientist-steward partnerships are increasingly common as scientists see the imperative for applying their knowledge to make a difference in local communities.

For scientists, research conducted in collaboration with civic ecology stewards is a means to move beyond the halls of academia and to act on their commitment to steer society toward a more resilient and sustainable future.[32] And for civic ecology stewards, partnering with scientists provides a means to measure their impacts and to adapt their practices accordingly. Adaptation to improve practices is one-half of a natural resource management approach referred to as adaptive co-management.

Graduate Student Scientists Supporting Civic Ecology Practices

At universities, the "skeptical scientists" are often graduate students, such as the students associated with our Civic Ecology Lab at Cornell who conduct participatory research in New York.

Megan Gregory worked with community gardeners to identify soil management as a critical concern, and then conducted participatory research with the gardeners on cover cropping as a means to improve their soils.[33]

Alex Kudryavtsev and Lilly Briggs worked with educators from community organizations that engage youth in civic ecology practices along the Bronx River—Alex to measure sense of place as an important outcome of after-school programs for youth[34] and Lilly to connect efforts to monitor changes in bird populations with a bird habitat restoration project.[35]

Philip Silva works with the Gowanus Canal Conservancy to monitor outcomes of street tree stewardship and stormwater management efforts.

And in the middle of Bryce DuBois's research on place attachment among surfers in the Rockaways, Hurricane Sandy barreled into his study sites. Among the first to respond were the surfers including Bryce—who established distribution centers for residents lacking food and water, and shortly thereafter, joined local residents to start rebuilding the dunes that would help protect their homes and beach in future storm surges.[36]

We take up adaptive co-management later in this chapter, but first we revisit the councilman who has recently left the gallery with a label for the collection of practices.

## How Do Civic Ecology Practices Expand and Spread?

Imagine that the graduate students, scientists, and civic ecology stewards have developed a research partnership at two sites along a daylighted river. They have collected data on the impact of two constructed wetlands on water and contaminant runoff into the river, and have adapted and strengthened their practices based on their findings. The councilman, seeing the impact of this partnership, has become interested in their efforts and wants to offer his support. Yet the partnership focuses on only two sites in a single city. What might the councilman do to expand the impacts of civic ecology from the local to the regional scale?

One approach is to tell others in his district to replicate the civic ecology practice. Replication is commonly tried with educational innovations—one school develops a successful model and attempts are made to replicate the same approach in other schools and in other cities. But because they are tied to particular places, civic ecology practices do not readily lend themselves to replication at other locations. Any one practice depends on the particular characteristics of its "place"—its history, environment, culture, demographics, and social-ecological memories. The nature of a particular practice also depends on local leadership, initiative, and participation. Attempts to impose a practice from the outside might act counter to the initiative of local leaders and activists. For these reasons, rather than spreading through an intentional process of replication, civic ecology initiatives often "catch on" through a more organic process of local leaders hearing about and being inspired by other practices, and then adapting the practices to fit local places and social-ecological contexts.[37]

An example of inspiration and adaptation comes from how the elevated High Line park in New York's Meatpacking District has led to similar efforts in other cities, each trying to create a version of the original idea that reflects local place. According to a *New York Times* article entitled "After High Line's Success, Other Cities Look Up":

Phone calls and visitors and, yes, dreams from around the world are pouring into the small offices of the Friends of the High Line on West 20th Street in Manhattan these days.

Detroit is thinking big about an abandoned train station. Jersey City and Philadelphia have defunct railroad beds, and Chicago has old train tracks that don't look like much now, but maybe they too . . .

The High Line's success as an elevated park, its improbable evolution from old trestle into glittering urban amenity, has motivated a whole host of public officials and city planners to consider or revisit efforts to convert relics from their own industrial pasts into potential economic engines.

In many of these places there had already been some talk and visions of what might be, but now New York's accomplishment is providing ammunition for boosters while giving skeptics much-needed evidence of the potential for success. The High Line has become, like bagels and CompStat, another kind of New York export.

"There's a nice healthy competition between big American cities," said Ben Helphand, who is pushing to create a park on a defunct rail line in Chicago. "That this has been done in New York puts the onus on us to do it ourselves and to give it a Chicago stamp."[38]

The High Line park in New York has inspired other cities to dream of repurposing postindustrial relics. In Philadelphia, Friends of the Rail Park hopes to transform the defunct Reading Viaduct into a public park and trail. Photo credit: Annik LaFarge

As civic ecology practices increasingly gain recognition and are covered in the media, we can expect them to spread through people being inspired by what others are doing, and implementing similar practices but with a local imprint. But the councilman may want to do more than simply wait for inspiring stories and the spread of practices on their own. He might have observed that civic ecology initiatives sometimes partner with other organizations and wonder how this happens.

### Adaptive Co-management

Adaptive co-management is an approach to environmental management that captures adaptation and collaboration.[39] Friends of Mitchell Park in Washington, DC, adapted to a drought by irrigating trees and monitoring tree mortality. They were then able to collaborate with the city Department of Parks and Recreation and a private sector tree care company to replace the dead trees and ensure that surviving trees remained healthy.

One way of envisioning adaptive co-management is to imagine two intersecting feedback loops. The first "adaptation loop" entails monitoring existing environmental and social conditions, implementing small interventions or management experiments (for example, adding lime to reduce acid levels in a lake), monitoring the changes as a result of the experiments, and using the information to inform future interventions. This loop leads to ongoing improvement of the stewardship or management practices so that they reflect data gathered about outcomes. It falls under a type of loop referred to as "information feedback" and makes a single practice more effective.

The second "collaboration loop" entails groups coming together to engage in hands-on stewardship, exchanging knowledge and perspectives, and building social connections and trust. The relationships and trust enable further work together and facilitate adding new nonprofit and government partners. The collaboration loop enables the practice to become part of a regional natural resources management system. It falls under the category of a "positive feedback loop," in that each activity reinforces the other.

Adaptive co-management thus links adaptation to strengthen the practice with collaboration to expand the practice. It is also a means of transforming a small, local practice—what we have referred to as a "social-ecological innovation"—into a regional management scheme or "policy innovation."[40]

### Policy Entrepreneur as Environment Shaper

Partnerships with scientists are critically important to the adaptation loop. But who can help civic ecology practices link across multiple stewardship practices and organizations in the collaboration loop?

Having exited the photo gallery, and having understood how civic ecology practices can be strengthened and contribute to environmental management at scales larger than a single lake or vacant lot, the councilman might be thinking through existing city priorities. Does current open space policy balance economic, civic, and environmental interests? How can he shape the legal and management environment so as to promote and leverage civic ecology practices, and in to doing support the city's agenda?

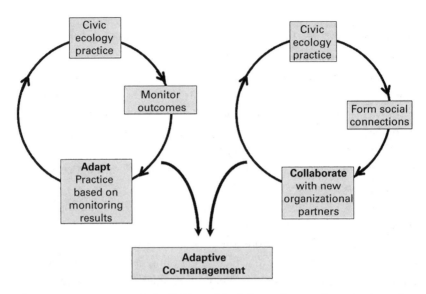

Adaptive co-management is the linking of the information or adaptation loop that uses monitoring to improve a practice, with the positive feedback loop of expanding connections and collaboration that enables civic ecology practices to scale out. Together the two loops broaden the impacts of civic ecology practice.

One way to understand how future policies might impact civic ecology practices is to look at the past. The history of community gardens in New York is a history of how policymakers changed their perception of civic ecology practices and then shaped the environment to support them.

The community gardening movement in New York began in broken neighborhoods not unlike the broken places we have visited throughout this book. In the 1970s, local activists calling themselves Green Guerillas, no longer able to stomach the drug use, rubble, and stench of decaying trash, began lobbing "seed bombs" packed with seeds, fertilizer, and water over fences into abandoned lots. The seeds took root, and as others recognized the power of these self-organized cleanup efforts, trash dumping declined and greening spread throughout New York neighborhoods. In the late 1970s, the city decided to support the community gardeners, by offering token leases for as little as $1 per year, and by creating the GreenThumb program to provide technical advice, compost, and fences.

Cooperation between the gardeners and the municipality blossomed through the late 1990s. But by that time the city had acquired many "vacant" lots due to delinquent tax bills, and real estate values were booming. Mayor Giuliani decided to sell off city-owned vacant properties to promote economic development. Whereas many of the properties were truly vacant, over one hundred were occupied by community gardens. From the mayor's perspective, the community gardens did not contribute to the tax rolls. But to the gardeners and their neighbors, the gardens were invaluable. The gardeners protested, defying the city and its bulldozers by chaining themselves to trees and fences in the gardens. Finally, the state attorney general granted a stay, which led to transfer to the parks department for some of the threatened community gardens. Joining the fray, the nonprofit Trust for Public Land hooked up with celebrity entertainer Bette Midler and her new nonprofit New York Restoration Project, to buy up community gardens not included in the city parks program. Shortly thereafter, three newly formed urban land trusts granted permanent tenure to gardens not under the parks department.[41]

Although overall the government and nonprofit support has helped community gardens in New York to persist and to flourish, some fear that gardens designated as city parks have taken on a new, less welcoming "feel" as they became more formal green spaces with locked gates and posted visiting hours. Others fear that community gardens, by making neighborhoods more pleasant and healthy, are contributing to a new cycle of economic development and gentrification. Despite these valid concerns, the community gardening case suggests that governments granting land tenure and implementing policies that provide the legal framework for civic ecology practices to exist and thrive is an important step in shaping a positive environment for stewardship practices.[42]

### Policy Entrepreneur as Scale-Crossing Broker

Our city councilman by now has become a policy entrepreneur. He has recognized the community initiative inherent to civic ecology practices, thrown his support behind community organizations, and helped to shape the legal and policy environment so as to facilitate the growth and spread of the practices in the gallery.[43] But he realizes that the small scales at which these practices operate—a single lot, a forest patch, or a small stretch of river—do not overlap with the scales at which government agencies, such as parks departments, manage the landscape.[44] Thus an important component of "shaping the environment" to enable civic ecology practices to flourish is networking across different practices and organizations operating at different scales.[45] One way to do this is by supporting individuals and organizations who "broker" information across different levels of organization—from allotment gardens, pocket parks, or artificial oyster reefs, to a citywide system of open space and watershed management.

"Scale-crossing brokers" integrate civic ecology practices into regional adaptive co-management systems. In order to keep the adaptation and collaboration feedback cycles on a positive trajectory, the scale-crossing broker needs to be vigilant in her observations of how the world around the practices is changing, as well as of change within the practices. She must stay one step ahead of the process—continually anticipating and responding to change by crafting opportunities for experimentation and learning, and stimulating new ways of thinking about problems and policy options. One challenge for a policy entrepreneur is to avoid forcing her own ideas and thus abandoning a more collaborative process. By meeting this challenge, policy entrepreneurs help prevent themselves— and the larger social-ecological systems in which they work—from falling into rigidity traps, repeating the same thing over and over and unable to respond to ongoing, inevitable, and perhaps ever greater change and upheavals.[46]

In the case of the Bronx River watershed, city government recognized the capacity of a community organization to serve as a scale-crossing broker. In granting the lead role to an organization that emerged out of community coalitions bent on restoring the river, the municipal government also recognized the self-organized nature of civic ecology practices.

The community bordering the Bronx River just before it flows into the East River near LaGuardia Airport is marked by high rises, warehouses, bodegas, and paved-over lots. For over a century, it has been plagued by industrial pollution, sewage, and dumping. In the 1970s, you could cross the river simply by skipping from one dumped refrigerator to the next. The *Yonkers Herald* described the river as an open sewer whose stench accosted passersby long before they even caught sight of the water.[47] Burnt-out apartment buildings and rampant crime made the Bronx a symbol of poverty and failing U.S. cities. Yet as people who had the means fled the Bronx, a small group of community activists took it upon themselves to bring back what had been lost. In 1974, Ruth Anderberg quit her job as a secretary at Fordham University and founded the Bronx River Restoration Project, spearheading efforts to remove seventy thousand tires from the river bottom.[48] By the 1980s, youth groups were working with the Bronx River Restoration Project and the city parks department, digging through hard packed mud to uncover a long-neglected twelve-foot wide path along the river, cutting back brush and weeds, and clearing the litter from the trail.[49] These initial efforts not only led to an expanding cycle of restoration activity on the part of community groups, but also were recognized with supportive action from federal, state, and city government agencies.[50]

In 2001, recognizing the role of civil society actors in spurring and coordinating local stewardship action, the City of New York threw its support behind the nonprofit Bronx River Alliance as the lead organization for the ever-expanding restoration and stewardship initiatives. The Alliance serves as a bridging organization that links the work of civic ecology stewards, scientists, government planners, regulatory agencies, designers, and advocates in recreating the Bronx River Greenway. The Greenway in turn has been recognized as an important community resource and has received $11 million in funding from the federal government. More recently, the Alliance partnered with Bronx community organizations to transform a former concrete factory to Concrete Plant Park.

Today the Bronx River Alliance continues to serve in a broker role, advocating for green space, promoting boating and other forms of recreation along the river, and supporting partnerships between community organizations and scientists to restore fish, oysters, and other species in the river, as well as vegetation along its banks. Although the image of the Bronx River as a stinking sewer and dumping ground originally spurred these efforts, today other symbols have emerged that are attracting interest and helping to sustain the Bronx River stewardship coalition. In addition to the return of the alewife (fish), symbols of a healthier river include a solitary beaver affectionately nicknamed "José," the first of its kind spotted after a two-hundred-year absence. The beaver takes his name from U.S. Representative José E. Serrano of the Bronx, himself a policy entrepreneur who has directed $15 million in federal funds toward the river's rebirth.[51] Today, the Bronx River Alliance works in close partnership with over sixty community and nonprofit organizations; city, state, and federal government agencies; schools; and businesses to serve as "a coordinated voice for the river."[52]

## Bridging across Scales and Places

A river connects residents living along its banks and within its watershed. The Bronx River not only connected people physically, but also provided a collective vision of how residents could transform a "burnt-out dumping ground." The Bronx River Alliance could claim that it "works in harmonious partnership to protect, improve and restore the Bronx River corridor so that it can be a healthy ecological, recreational, educational and economic resource for the *communities through which the river flows.*"[53]

But what about civic ecology groups stewarding other natural resources, such as community gardens, which are dispersed more or less randomly across the city with no river or other physical feature to connect people across sites? Are there ways to bridge across such practices?

In addition to focusing on areas that are connected physically by a river, bridging organizations have linked stewardship practices that occur in seemingly disconnected patches, such as city parks and community gardens. This is the case in Seattle, Washington, where volunteer friends groups have coalesced around individual city parks, and the nonprofit organization EarthCorps serves as a scale-crossing broker through sharing information using its citywide GIS mapping of parks and vegetation. On a regional scale also in the Pacific Northwest, the nonprofit ForTerra links conservation efforts in the iconic national parks that are home to the majestic Cascade and Olympic Mountains with their glacier-strewn peaks and towering Douglas fir and western red cedar forests, with conservation actions in Seattle's and Tacoma's urban community gardens.[54]

Sometimes civic ecology organizations are connected by their belief in a cause rather than a physical resource such as a river or a region like the Pacific Northwest. The American Community Gardening Association serves as a broker for individual community gardens committed to a broad vision of greening and growing food.[55] And the U.S. EPA's Urban Waters Federal Partnership bridges eleven federal agencies and hundreds of community groups to "revitalize urban waters and the communities that surround them, transforming overlooked assets into treasured centerpieces and drivers of urban revival."[56] Finally, moving up from the local to the regional, national, and finally global scale, the international-al Urban Biosphere Initiative (URBIS) serves as a scale-crossing broker across numerous organizations and city governments from eighty-four countries around the world—including those engaged in civic ecology stewardship—to harness cities' "unprecedented and often untapped opportunities for innovation" and "to reconcile urban development with the conservation of biodiversity and the sustainable use of natural resources—a quest to engender cities with greater social-ecological resilience in the context of global environmental change."[57]

In short, policy entrepreneurs—including environment shapers and bridging organizations like city governments and initiatives like URBIS—operate at local, regional, national, and even global scales to facilitate the exchange of ideas and practice as well as to mobilize groups to adapt practices for their local contexts.[58] Further, they expand the impact of practices through linking them to government, nonprofit, and scientific organizations with resources not available to individual civic ecology practices.[59] In facilitating such networks, policy entrepreneurs play a critical role in expanding cycles of information exchange and collaboration that transform civic ecology practices—local, self-organized initiatives focused on single resources—into regional adaptive co-management systems. In some cases, these regional management systems grow into a national movement or even become part of international agreements on biodiversity and ecosystem services.

Our city councilman, having seen the gallery of civic ecology practices and their results, and having helped to shape the legal and organizational environment, can support such a process of bridging across individual stewardship efforts to form a citywide network of groups committed to resilience and sustainability. And he can partner with others to scale out the practices in his city so that they become part of regional, national, and even global resilience and sustainability networks.

## Summing Up

In previous chapters, we have depicted civic ecology practices as virtuous cycles of greening, ecosystem services provision, and civic renewal.[60] We also have shown how such cycles can be embedded in the rebuilding and other phases of the adaptive cycle, and as part of larger panarchies of adaptive cycles operating at different scales. This chapter suggests additional cycles of building regional capacity through experimentation and adaptation, and through networking and collaboration. Civic ecology stewards who form partnerships with scientists to monitor their impacts can use the data and their newly developed research capacity to adapt their practices on an ongoing basis, a process that we have described as the information or adaptation cycle. This cycle in turn interacts with the collaboration cycle that involves sharing ideas and building trust across civic ecology practices and with nonprofit and government actors. Together the adaptation cycle strengthens individual practices and the collaboration cycle helps to scale out local efforts to form regional adaptive co-management systems. Although we have separated internal capacity building within civic ecology practices from networking across scales and practices in our treatment, processes at different levels invariably interact. A government initiative that helps individual practices monitor their impacts also may provide opportunities for information sharing across multiple practices and organizations.

Locating virtuous cycles of greening, ecosystem services provision, and building social capital is an important task for a policy entrepreneur. The city councilman or manager of a citywide park system who is able to locate these virtuous cycles in his community and create an environment that further empowers the stewards in these cycles, is helping to shape the environment so that it supports and expands the impacts of civic ecology practices.[61] Once the virtuous cycles are identified, they can be managed by individuals and organizations that serve as scale-crossing brokers. The networks created by brokers lead to additional virtuous cycles of increasing numbers of organizations coming together to manage landscapes. This in turn attracts the attention of policy entrepreneurs working at regional and even global scales.

A critical lesson from the Bronx River Alliance and similar efforts is that government can provide financial and technical support, as well as enabling legislation that grants land tenure and management rights to civic ecology practices, while allowing the nonprofit and community organizations to play the lead role. In this way, the efforts of the more flexible nonprofit organization with a history of self-organized stewardship—the "bees" who "have the new ideas, and are mobile, quick and able to cross-pollinate"—are leveraged rather than constrained or even suppressed by the larger organizations—the "trees" with deeper roots.[62] In short, the challenge for policy entrepreneurs is to identify existing initiatives that foster social capital, place-based stewardship, and virtuous cycles of greening and civic renewal; to build the capacity of the leaders and participants through providing secure land tenure, learning opportunities including those encompassing participatory research and monitoring, and incentives linked not to personal aims but rather to the production of public goods; and finally to help civic ecology practices and organizations at multiple levels join together to form regional systems of adaptive co-management and global networks of information sharing and recognition.

# Afterword: Moving Forward

On October 22, 2012, Hurricane Sandy struck New York. As water rushed into the World Trade Center site, New York State Governor Andrew Cuomo descended seven stories into the Ground Zero pit. He wanted to be sure the water was flowing from the Hudson River and not over a breach in the retaining wall intended to protect the construction. A day later, Cuomo was on the phone with President Obama. Recalling Hurricane Irene just fourteen months earlier, Cuomo said: "we have a 100-year flood every two years now." Reflecting his anxiety and a sense of unease about the future, Cuomo contradicted himself: "There's no such thing as a 100-year flood. These are extreme weather patterns. The frequency has been increasing."[1]

Immediately after the waters receded revealing the devastation in New York, the response was not unlike that after the 9/11 terrorist attacks eleven years earlier. New Yorkers were looking to install the wall that would thwart the next disaster—only this time instead of politicians talking about a wall of security, engineers envisioned massive seawalls.[2]

Yet, just as New Yorkers turned to planting as a way to memorialize, restore, and recover following 9/11, less than a month after Hurricane Sandy, volunteer efforts—including civic ecology practices—were starting to emerge. The Bronx River Alliance called for volunteers to look toward the future by replacing trees lost in the storm:

Join us on November 10 for our rescheduled planting of 500 trees at Shoelace Park—in the wake of Hurricane Sandy be a part of the next generation of trees.[3]

And people were recognizing how perhaps instead of a seawall, a civic ecology approach might help to buffer the next storm. *New York Times* Dot Earth columnist Andrew Revkin asked readers to consider oyster restoration efforts, such as those at Governors' Island Harbor School, as a means to protect the shoreline.[4] And residents in Long Beach, NY, were coming together to heal themselves, their community, and their dunes through constructing a storm barrier of discarded Christmas trees arranged along the beach.[5]

Climate change and its companions—rising sea level and flooding—loom over coastal cities like a colossal darkening cloud. What role will tiny efforts—like restoring an oyster reef in the New York estuary—play in the "new normal" of climate change? Will these efforts, like those of the girl who risked her life to find a single leaf for her dying sister in the World War II Warsaw ghetto, provide only momentary happiness followed by death—by the demise of our everyday routines, our social-ecological systems as we know them?

We struggle to put in place national and multilateral agreements and we consider the billions in costs to build protective seawalls. Yet alongside these and other efforts, there are reasons to consider a role for local initiatives including civic ecology practices. Civic ecology stewards already have developed models for how communities adapt and transform when faced with violence, poverty, disinvestment, environmental contamination, and myriad other stresses, and thus their work offers lessons for dealing with new stresses offered up by climate change. Their practices provide a model for linking people concerned about the community with those concerned about the environment—integrating social and ecological solutions is essential to addressing social-ecological dilemmas like climate change. By linking up with each other, with scientists and with policymakers, civic ecology stewards can grow their practices into regional initiatives that have the potential to mitigate and to help society to adapt. And when things get really bad, perhaps their memories and experience in transforming communities and nature can be used to transform broken places into places that are more nourishing of the human spirit and the life that surrounds and nourishes us.

Through touching, hearing, and smelling nature, civic ecology stewards experience the restorative power of nature and the "spats" of hope that keep people from delving into apathy and despair. In this sense, civic ecology presents a more hopeful—and perhaps realistic—view of what is needed in order for people to "live well in nature"[6] Here we return to the words of Arne Næss:

We need environmental ethics, but when people feel that they unselfishly give up, or even sacrifice, their self-interests to show love for nature, this is probably, in the long run, a treacherous basis for conservation. Through identification, they may come to see that their own interests are served through conservation, through genuine self-love, the love of a widened and deepened self.[7]

It is in these realities that the greatest promise of civic ecology emerges. Our species successfully evolved and adapted to millions of years of shocks and disturbances, all the while understanding that human issues are embedded in ecological issues. But humans have experienced a collective amnesia, where we demonstrate daily a profound forgetting of our fundamental, evolutionary roots as part of a larger group of organisms and interactions we call nature. Civic ecology practices are hopeful indications of returning instincts, of restored memories of our ecological selves and ecological home, of the very stuff we are made of. The ecosystem we refer to, and remember through these actions, is a part of us, and we a part of it. We are inextricably bound together, despite the increasingly deafening drumbeat of those who would deceive us further into believing that we can escape from the proverbial garden, that we are ascendant, that we have experienced a transfiguration and escaped our biology. Fleeing from our biology, and attempting to obliterate its memory in our technological advances, may underlie many of the broken places we see speak of, dotted across the landscape. But like hopeful prodigals, we are returning, we are remembering. This is the hope offered forth by the unbroken spirits defying broken places in civic ecology practices.

# Notes

## Preface

1. Alex Kudryavtsev first suggested we use the term "broken places." See also A. C. Madrigal, "The Sacrificial Landscape of *True Detective*," *The Atlantic*, March 7, 2014, http://www.theatlantic.com/technology/archive/2014/03/the-sacrifical-landscape-of-em-true-detective-em/284302/.

2. E. Moser, "Seeding the City," n.d., http://www.evemosher.com/2011/seedingthecity/.

3. People who study civic ecology are called "civic ecologists." Community leaders and volunteers who conduct civic ecology practices are called "civic ecology practitioners" or "civic ecology stewards."

4. Disinvestment has been described as "an economic disease that can ravage cities like an aggressive cancer" or "the process in which businesses elect not to invest further capital in their ongoing operations." Further, "decline in investment results in a decline in the quality of the assets that remain, and that further feeds the logic that additional investment is futile, which in turn makes the downward spiral more severe." M. Pollock, "Cut Casino Tax Rate to Encourage Investment," *The Press of Atlantic City*, 2014, http://www.pressofatlanticcity.com/opin ion/commentary/michael-pollock-cut-casino-tax-rate-to-encourage-investment/article_8373532b-d200-5877-8b60-ee364b1ed920.html?mode=jqm. See also chapter 9 on vicious cycles.

5. A. de Tocqueville, *Democracy in America* (Cambridge, MA: Sever & Francis Press, 1863).

6. R. B. Putnam, Bowling Alone: The Collapse and Revival of American Community (New York: Simon & Schuster, 2000).

7. C. Sirianni and L. A. Friedland, *The Civic Renewal Movement: Community Building and Democracy in the United States* (Dayton, OH: Charles F. Kettering Foundation, 2005).

8. A. Leopold, *A Sand County Almanac* (New York: Oxford University Press, 1949).

9. Ibid., 203.

10. Lewis Friedland describes a community's civic ecology as "the entirety of civic relationships in a defined area," or "the overall environment in which young people are socialized into civic life." L. A. Friedland and S. Morimoto, *The Changing Lifeworld of Young People: Risk, Resume-Padding, and Civic Engagement*, The Center for Information & Research on Civic Learning and Engagement (CIRCLE) (College Park, MD: CIRCLE, 2005). See also J. V. Riker and K. E. Nelson, "What Works to Strengthen Civic Engagement in America: A Guide to Local Action and Civic Innovation," (College Park, MD: The Democracy Collaborative-Knight Foundation Civic Engagement Project, 2003).

11. T. W. Smith, "Civic Ecology: A Community Systems Approach to Sustainability," *Oregon Planners' Journal* (March/April 2008): 10–12, 14–15.

12. K. Poole, "Civitas Oecologie: Infrastructure in the Ecological City," in *The Harvard Architect Review*, ed. T. Genovese, L. Eastley, and D. Snyder (New York: Princeton Architectural Press, 1998), 126–142.

13. K. L. Wolf, "Human Dimensions of Urban Forestry and Urban Greening," http://www.naturewithin.info/civic.html.

14. K. L. Wolf et al., "Environmental Stewardship Footprint Research: Linking Human Agency and Ecosystem Health in the Puget Sound Region," *Urban Ecosystems*, April 26, 2011, http://www.naturewithin.info/CivicEco/UrbanEcosystems.Stewardship.Apr2011online.pdf; E. S. Svendsen and L. Campbell, "Urban Ecological Stewardship: Understanding the Structure, Function and Network of Community-Based Land Management," *Cities and the Environment* 1, no. 1 (2008), http://digitalcommons.lmu.edu/cate/vol1/iss1/4/.

15. K. G. Tidball and M. E. Krasny, "From Risk to Resilience: What Role for Community Greening and Civic Ecology in Cities?" in *Social Learning Toward a More Sustainable World*, ed. A. E. J. Wals, 149–164 (Wagengingen, The Netherlands: Wagengingen Academic Press, 2007).

16. R. J. Borden, "A Brief History of SHE: Reflections on the Founding and First Twenty-Five Years of the Society for Human Ecology," *Human Ecology Review* 15, no. 1 (2008).

17. D. P. Robertson and R. B. Hull, "Public Ecology: An Environmental Science and Policy for Global Society," *Environmental Science & Policy* 6 (2003): 399–410.

18. M. Bookchin, "What Is Social Ecology?," in *Environmental Philosophy: from Animal Rights to Radical Ecology*, ed. M. E. Zimmerman, J. B. Callicott, G. Sessions, K. J. Warren, and J. Clark (Englewood Cliffs, NJ: Prentice Hall, 1993), 354–373.

19. D. Faber and D. McCarthy, "The Evolving Structure of the Environmental Justice Movement in the United States: New Models for Democratic Decision-Making," *Social Justice Research* 14, no. 4 (2001): 405–421; C. R. Palamar, "The Justice of Ecological Restoration: Environmental History, Health, Ecology, and Justice in the United States," *Human Ecology Review* 15, no. 1 (2008): 82–94.

20. T. M. Schusler, D. J. Decker, and M. J. Pfeffer, "Social Learning for Collaborative Natural Resource Management," *Society & Natural Resources* 15 (2003):

309–326; D. Schlosberg and J. S. Dryzek, "Political Strategies of American Environmentalism: Inclusion and Beyond," *Society & Natural Resources* 15 (2002): 787–804.

21. W. Burch and J. M. Grove, "Ecosystem Management: Some Social, Conceptual, Scientific, and Operational Guidelines for Practitioners," in *Ecological Stewardship: A Common Reference for Ecosystem Management*, ed. N. C. Johnson et al., 279–295 (Oxford: Elsevier Science Ltd., 1999).

22. F. Earls and M. Carlson, "The Social Ecology of Child Health and Well-Being," *Annual Review of Public Health* 22 (2001): 143–166.

23. E. T. Wimberley, *Nested Ecology: The Place of Humans in the Ecological Hierarchy* (Baltimore: The Johns Hopkins University Press, 2009).

24. E. Ostrom, "Polycentric Systems for Coping with Collective Action and Global Environmental Change," *Global Environmental Change* 20 (2010) 550–557; E. Ostrom, *Governing the Commons* (New York: Cambridge University Press, 1990); T. Dietz, E. Ostrom, and P. C. Stern, "The Struggle to Govern the Commons," *Science* 302, no. 5652 (2003): 1907–1912.

25. C. Folke et al., "Resilience and Sustainable Development: Building Adaptive Capacity in a World of Transformations," World Summit on Sustainable Development, Johannesburg, South Africa, April 16, 2002, http://www.resalliance.org/files/1144440669_resilience_and_sustainable_development.pdf.

26. In addition to environmental movement, civic environmentalism is a form of governance and an innovative approach to civic renewal. D. John, "Civic Environmentalism," in *Environmental Governance Reconsidered*, ed. R. F. Durant, D. Fiorini, and R. O'Leary, 219–254 (Cambridge, MA: MIT Press, 2004); C. Sirianni, *Investing in Democracy: Engaging Citizens in Collaborative Governance* (Washington, DC: Brookings Institution Press, 2009); C. Sirianni and L. A. Friedland, *Civic Innovation in America: Community Empowerment, Public Policy, and the Movement for Civic Renewal* (Berkeley: University of California Press, 2001); Sirianni and Friedland, *Civic Renewal Movement*. The philosopher Andrew Light uses civic environmentalism and urban ecological citizenship to describe a new urban environmentalism that integrates civic and environmental values; see Light, "Urban Ecological Citizenship," *Journal of Social Philosophy* 34, no. 1 (2003): 44–63.

27. J. L. Dickinson and R. Bonney, eds., *Citizen Science: Public Collaboration in Environmental Research* (Ithaca, NY: Cornell University Press, 2012); R. Bonney et al., "Citizen Science: A Developing Tool for Expanding Science Knowledge and Scientific Literacy," *BioScience* 59, no. 11 (2009): 977–984.

28. Some citizen science projects integrate data collection with stewardship, such as planting flowers that attract pollinators. Great Sunflower Project, n.d., http://www.greatsunflower.org/; Monarch Watch, n.d., http://www.monarchwatch.org/; L. Briggs and M. E. Krasny, "Conservation in Cities: Linking Citizen Science and Civic Ecology Practices" (Ithaca, NY: Civic Ecology Lab, Cornell University, 2013), http://civeco.files.wordpress.com/2013/09/2013-briggs.pdf.

29. R. D. Bullard and G. S. Johnson, "Environmental Justice: Grassroots Activism and Its Impact on Public Policy Decision Making," *Journal of Social Issues* 56, no. 3 (2000): 555–578.

30. S. R. Kellert, J. Heerwagen, and M. Mador, eds., *Biophilic Design: The Theory, Science, and Practice of Bringing Buildings to Life* (Hoboken, NJ: John Wiley & Sons, 2008).

31. T. Beatley, Biophilic Cities: Integrating Nature into Urban Design and Planning (Washington, DC: Island Press, 2011).

32. See forward to this volume by Robert Gottlieb.

33. B. DuBois and K. G. Tidball, "Greening in Coastal New York Post-Sandy," Civic Ecology Lab White Paper Series (Ithaca, NY: Cornell Civic Ecology Lab, 2013). Other examples of combining recreation with civic ecology stewardship come from Marianne Krasny's work with Friends of the Gorge, a Cornell University student group that integrates hiking with gorge cleanups and trail restoration; M. E. Krasny and J. Delia, "Natural Area Stewardship as Part of Campus Sustainability," *Journal of Cleaner Production*, May 9, 2014, http://www.sciencedirect .com/science/article/pii/S0959652614003667.

Similarly, Keith Tidball works with military veteran hunters, anglers, and other types of outdoor enthusiasts, who help restore wetlands for waterfowl, streams for trout and salmon, and forests for wildlife; K. G. Tidball, "Outdoor Recreation, Restoration and Healing for Returning Combatants," September 23, 2013, http:// www.thenatureofcities.com/2013/09/23/outdoor-recreation-restoration-and -healing-for-returning-combatants/.

The case of the surfers in New York also illustrates how recreationists often turn to civic ecology practices after some kind of hardship or disaster. One explanation is that through recreation, surfers and other recreationists are able to express their affinity for life and at the same time develop an attachment to, or love of, a particular place. When that place is threatened by disaster, they return in an attempt to "reconcile" their relationship with a nature that seemingly has turned on them and to express their grief. They then feel a need to restore—to try to bring back—the place they knew. Hence the surfers' engagement in dune restoration after Hurricane Sandy. See chapter 1 for a discussion of broken places, and chapter 2 for a discussion of biophilia, topophilia, disaster, and greening. See also R. C. Stedman and M. Ingalls, "Topophilia, Biophilia and Greening in the Red Zone," in *Greening in the Red Zone*, ed. K. G. Tidball and M. E. Krasny, 129–144 (New York: Springer, 2014).

34. NOAA, "Military Veterans Help Rebuild Fisheries," http://www.habitat .noaa.gov/highlights/vetsrestorationvideo.html. For returning veterans who have seen combat, fishing provides an opportunity to reconnect with life. As in the case of the surfers, recreation (fishing) can be an entry point for restoration of trout and salmon streams. It may also be a means of emotional restoration for the veterans. See chapter 6 for a discussion of therapeutic outcomes of spending time in and restoring nature.

35. Svendsen and Campbell, "Urban Ecological Stewardship"; Wolf et al., "Environmental Stewardship Footprint Research."

36. Other frameworks used in analyses of civic environmental stewardship include: governance, J. J. Connolly et al., "Organizing Urban Ecosystem Services through Environmental Stewardship Governance in New York City," *Landscape and Urban Planning* 109 (2013): 76–84; institutional analysis, D. R. Fisher, L. Campbell, and E. S. Svendsen, "The Organisational Structure of Urban Environmental Stewardship," *Environmental Politics* 21, no. 1 (2012): 26–48; and network analysis, Svendsen and Campbell, "Urban Ecological Stewardship."

37. H. Rolston, "Nature for Real: Is Nature a Social Construction?" in *The Philosophy of the Environment*, ed. T. D. J. Chappell (Edinburgh: University of Edinburgh Press, 1997), 39.

38. Rolston discusses at length the problematic use of the term "nature," given humans are part of nature, but also that "the word 'nature' arises in our language, constructed by humans, because we need a container matching this world that contains all these myriads of creatures and phenomena we encounter, lions and five million other species, and mountains, rivers and ecosystems" (ibid., 43).

Rolston also addresses the problematic use of the word "environment," which he defines similarly to our use of this term: "My environment when encountered as a landscape is a commons shared, your environment too, *our* environment. That fosters social solidarity, fortunately. It also demands another, fuller sense in which *the* environment is objectively out there, and this is not only our social world, but the natural world that we move through, there before we arrive, and there after we are gone" (ibid., 44; italics in original).

39. S. Grant and M. Humphries, "Critical Evaluation of Appreciative Inquiry: Bridging an Apparent Paradox," *Action Research* 4, no. 4 (2006) 401–418.

40. C. Nesper, "Legitimate Protection or Tactful Abandonment: Can Recent California Legislation Sustain the San Francisco Bay Area's Public Lands?," *Golden Gate University Environmental Law Journal* 6, no. 1 (2012): 153–181. See also R. I. Steinzor, "The Corruption of Civic Environmentalism," *The Environmental Law Reporter* 30, no. ELR 10909 (2000), http://elr.info/news-analysis/30/10909/corruption-civic-environmentalism.

41. J. Colding et al., "Urban Green Commons: Insights on Urban Common Property Systems," *Global Environmental Change* 23 (2013) 1039–1051; T. Slater, "The Resilience of Neoliberal Urbanism," January 28, 2014, http://www.opendemocracy.net/opensecurity/tom-slater/resilience-of-neoliberal-urbanism.

42. See M. J. Lee, "Possibilities for the Emergence of Civic Ecology Practices in Response to Social-Ecological Disturbance: The Case of Nuisance Chironomids in Singapore" (Ithaca, NY: Civic Ecology Lab, Cornell University, 2013), http://civeco.files.wordpress.com/2013/09/2013-lee.pdf.

## The Principled Chapter

1. This account is compiled from interviews with Dennis Chestnut conducted during 2013 and 2014, and from J. R. Wennersten, "The View from Watts Branch: Marvin Gaye Park Spurs an Environmental Transformation," http://groundworkdc.org/wp-content/uploads/2010/08/Watts-Branch-Restoration_May09.pdf; P.

Sullivan, "Getting Its Groove Back," *Washington Post*, April 2, 2006, http://www.washingtonpost.com/wp-dyn/content/article/2006/04/01/AR2006040101105.html; National Recreation and Park Foundation, "Marvin Gaye Park Backgrounder," http://www.nrpa.org/uploadedFiles/Events/MGP Backgrounder.pdf? n=1508.

2. Sullivan, "Getting Its Groove Back."

3. Center for Biological Diversity, "Bald Eagle Population Exceeds 11,000 Pairs in 2007," n.d., http://www.biologicaldiversity.org/species/birds/bald_eagle/report/; National Wildlife Federation, "Earth Conservation Corps (ECC): Reclaiming Two of America's Most Endangered Resources—Our Youth and Our Environment," n.d., http://www.nwf.org/news-and-magazines/media-center/faces-of-nwf/ecc.aspx.

4. In addition to Earth Conservation Corps and Groundwork, the partners include at the federal level: Environmental Protection Agency, Housing and Urban Development, National Park Service, U.S. Fish and Wildlife Service, and the Departments of Labor, Energy, Education, and Agriculture; in Washington, DC: Health Department, Parks and Recreation Department, District of Columbia Housing Authority, University of the District of Columbia; nonprofits: the Anacostia Waterfront Corporation, DC Promise Neighborhood, Temple of Praise, Anacostia Watershed Society, Washington Parks and People, Sierra Club, Casey Trees Foundation, and National Recreation and Park Association; and several businesses: Playworld Systems, Playcore, Landscape Structures, and Kompan; National Recreation and Park Foundation, "Marvin Gaye Park Backgrounder," http://www.nrpa.org/uploadedFiles/Events/MGP%20Backgrounder.pdf?n=1508.

5. The use of the term "hone in" as opposed to the more commonly used "home in" is the subject of a lively linguistic debate. During the 1980 presidential campaign, the "first" George Bush created a stir when he talked of "honing in on the issues," but since that time "hone in on" is gaining ground in popular usage. We prefer "hone in" because of its suggestion that we are sharpening something—in this case our understanding of civic ecology practices. See http://grammar.about.com/od/alightersideofwriting/a/homehonegloss.htm.

6. The principles in this chapter are adapted from M. E. Krasny and K. G. Tidball, "Civic Ecology: A Pathway for Earth Stewardship in Cities," *Frontiers in Ecology and the Environment* 10, no. 5 (2012): 267–273.

7. R. C. Stedman and M. Ingalls, "Topophilia, Biophilia and Greening in the Red Zone," in *Greening in the Red Zone*, ed. K. G. Tidball and M. E. Krasny (New York: Springer, 2014), 129–144.

8. T. S. Eisenman, "Frederick Law Olmsted, Green Infrastructure, and the Evolving City," *Journal of Planning History* 12 (2013): 287–311.

9. S. R. Kellert, *Kinship to Mastery: Biophilia in Human Evolution and Development* (Washington, DC: Island Press, 1997), 8.

10. S. R. Kellert, *Birthright: People and Nature in the Modern World* (New Haven: Yale University Press, 2012), xiii.

11. S. R. Kellert and E. O. Wilson, eds., *The Biophilia Hypothesis* (Washington, DC: Island Press, 1993). E. O. Wilson, *Biophilia* (Cambridge, MA: Harvard University Press, 1984).

12. K. G. Tidball, "Urgent Biophilia: Human-Nature Interactions and Biological Attractions in Disaster Resilience," *Ecology and Society* 17, no. 2 (2012), http://www.ecologyandsociety.org/vol17/iss2/art5/.

13. Although we emphasize the role of community leaders such as Dennis Chestnut in initiating civic ecology practices, sometimes government employees start similar practices. City environmental engineer David Gabbard and city forester David Swenk shared a concern about phosphorus runoff and algal blooms in rivers flowing through Lexington, Kentucky, and started Reforest the Bluegrass to engage volunteers in planting trees to mitigate storm runoff. M. Buranen, "Reforest the Bluegrass," *Stormwater*, March 16, 2000, http://www.stormh2o.com/SW/articles/16503.aspx.

14. S. Barthel, C. Folke, and J. Colding, "Social-Ecological Memory in Urban Gardens: Retaining the Capacity for Management of Ecosystem Services," *Global Environmental Change* 20 (2010): 255–265.

15. V. Schaefer, "Alien Invasions, Ecological Restoration in Cities and the Loss of Ecological Memory," *Restoration Ecology* 17, no. 2 (2009): 171–176.

16. Barthel, Folke, and Colding, "Social-Ecological Memory."

17. I. Davidson-Hunt and F. Berkes, "Learning as You Journey: Anishinaabe Perception of Social-Ecological Environments and Adaptive Learning," *Conservation Ecology* 8, no. 1 (2003), http://www.consecol.org/vol8/iss1/art5/5; S. Barthel et al., "Urban Gardens: Pockets of Social-Ecological Memory," in *Greening in the Red Zone*, ed. K. G. Tidball and M. E. Krasny (New York: Springer, 2014), 145–158.

18. MEA, *Millennium Ecosystem Assessment: Ecosystems and Human Well-Being: Synthesis* (Washington, DC: Island Press, 2005).

19. M. E. Krasny et al., "Civic Ecology Practices: Participatory Approaches to Generating and Measuring Ecosystem Services in Cities," *Ecosystems Services* 7 (March 2014): 177–186.

20. R. Louv, *Last Child in the Woods: Saving Our Children from Nature-Deficit Disorder* (New York: Algonquin Books, 2006).

21. A. Bandura, "Perceived Self-Efficacy in Cognitive Development and Functioning," *Educational Psychologist* 28, no. 2 (1993): 117–148; R. J. Slater, "Urban Agriculture, Gender and Empowerment: An Alternative View," *Development Southern Africa* 18, no. 5 (2001): 635–650; J. Warburton and M. Gooch, "Stewardship Volunteering by Older Australians: The Generative Response," *Local Environment* 12, no. 1 (2007): 43–55.

22. Friends of the Rouge Watershed, n.d., http://www.frw.ca.

23. EPA, "Urban Waters Federal Partnership," June 13, 2014, http://www.urbanwaters.gov.

24. E. Svendsen and L. Campbell, "Land-Markings: 12 Journeys through 9/11 Living Memorials" (Newtown Square, PA: U.S. Department of Agriculture, Forest Service, Northern Research Station, 2006), http://www.livingmemorialsproject .net/DOWNLOADS/NRS-INF-1-06.pdf; E. S. Svendsen and L. Campbell, "Living Memorials Project: Year 1 Social and Site Assessment," U.S. Forest Service General Technical Report NE-333, 2005; E. S. Svendsen and L. Campbell, "Community-Based Memorials to September 11, 2001: Environmental Stewardship as Memory Work," in *Greening in the Red Zone*, ed. Tidball and Krasny, 339–355.

25. Farlex, *The Free Dictionary*, n.d., http://www.thefreedictionary.com/vicious +cycle.

26. C. Folke, "Resilience: The Emergence of a Perspective for Social-Ecological Systems Analyses," *Global Environmental Change* 16 (2006): 253–267; L. H. Gunderson and C. S. Holling, eds., *Panarchy: Understanding Transformations in Human and Natural Systems* (Washington, DC: Island Press, 2002).

27. B. H. Walker and D. Salt, *Resilience Thinking: Sustaining Ecosystems and People in a Changing World* (Washington, DC: Island Press, 2006).

28. H. Ernstson and T. Elmqvist, "Globalization, Urban Ecosystems and Social-Ecological Innovations: A Comparative Network Analytic Approach" (Stockholm, Sweden: Stockholm University, 2011).

29. J. L. McKnight and J. P. Kretzman, "Mapping Community Capacity" (Evanston, IL: Institute for Policy Research, 1996); K. G. Tidball and E. D. Weinstein, "Applying the Environment Shaping Methodology: Conceptual and Practical Challenges," *Journal of Intervention and Statebuilding* 5, no. 4 (2012): 369–394; E. D. Weinstein and K. G. Tidball, "Environment Shaping: An Alternative Approach to Development and Aid," *Journal of Intervention and Statebuilding* 1, no. 1 (2007): 67–85.

30. J. Long, "Detroit's Demise Sets up Rebirth from Grassroots," *Bloomberg Businessweek*, April 6, 2011, http://www.bloomberg.com/news/2011-04-06/ detroit-s-fall-sets-up-rebirth-from-grassroots-commentary-by-joshua-long.html.

31. For example, Mayor Bloomberg's MillionTreesNYC initiative has resulted in hundreds of thousands of trees being planted in parks, courtyards, and yards, and along residential streets. Mayor Bloomberg also has been a national and international leader in climate change and sustainability. PlaNYC, "MillionTreesNYC," NYC Parks and NY Restoration Project, n.d., http://www.milliontreesnyc.org/ html/home/home.shtml.

# 1   Broken Places

1. This account is compiled from observations during Marianne Krasny's visit to Detroit in 2010, and from Keith Tidball's reflections, having spent his adolescence (1982–1989) living in the Downriver area of Detroit, as well as research conducted with Greening of Detroit.

2. Y. Allweil, "Shrinking Cities: Like a Slow-Motion Katrina," *Forum of Design for the Public Realm* 19, no. 1 (2007): 91–93, http://places.designobserver.com/media/pdf/Dispatches__--_456.pdf.

3. Similar to how scientists talk about thresholds, popular writer Malcolm Gladwell talks about tipping points. But his focus is not on system collapse, chaos, and falling into a new, undesirable state (like a eutrophic lake), but rather on how new ideas can slowly chug along and then, suddenly, spread like viruses. He defines tipping points as, "that magic moment when an idea, trend, or social behavior crosses a threshold, tips, and spreads like wildfire." M. Gladwell, *The Tipping Point: How Little Things Can Make a Big Difference* (New York: Little, Brown and Company, 2002). For a scientific treatment of thresholds and "planetary boundaries," see J. Rockström et al., "Planetary Boundaries: Exploring the Safe Operating Space for Humanity," *Ecology and Society* 14, no. 2 (2009), http://www.ecologyandsociety.org/vol14/iss2/art32/.

4. S. R. Carpenter, D. Ludwig, and W. A. Brock, "Management of Eutrophication for Lakes Subject to Potentially Irreversible Change," *Ecological Applications* 9, no. 3 (1999): 751–771.

5. B. J. McCay and A. C. Finlayson, "The Political Ecology of Crisis and Institutional Change: The Case of the Northern Cod," paper presented at the Annual Meeting of the American Anthropological Association, Washington, DC, 1995.

6. Rockström et al., "Planetary Boundaries." See also J. Hansen et al., "Assessing 'Dangerous Climate Change': Required Reduction of Carbon Emissions to Protect Young People, Future Generations and Nature," *PLoS ONE* 8, no. 12 (2013): e81648, http://dx.doi.org/10.1371%2Fjournal.pone.0081648.

7. M. Pelling and K. Dill, "Disaster Politics: Tipping Points for Change in the Adaptation of Sociopolitical Regimes," *Progress in Human Geography* 34, no. 1 (2010): 21–37; M. Pelling and D. Manuel-Navarrete, "From Resilience to Transformation: The Adaptive Cycle in Two Mexican Urban Centers," *Ecology and Society* 16, no. 2 (2011), http://www.ecologyandsociety.org/vol16/iss2/art11/; Gunderson and Holling, *Panarchy*.

8. E. D. Gonzalez, *The Bronx* (New York: Columbia University Press, 2004).

9. W. A. Shutkin, *The Land That Could Be: Environmentalism and Democracy in the Twenty-First Century* (Cambridge, MA: MIT Press, 2001); P. Medlar and H. Sklar, *Streets of Hope: The Fall and Rise of an Urban Neighborhood* (Boston: South End Press, 1994).

10. Medlar and Sklar, *Streets of Hope*, 5–6.

11. M. Francis, L. Cashdan, and L. Paxson, *Community Open Spaces* (Washington, DC: Island Press, 1984).

12. We recognize that not everyone considers the High Line favorably due to the "unsustainable" inputs of water and nutrients needed to maintain its plantings, and that the plant communities that had colonized the abandoned railway were destroyed. Although the plants occupying the railway included non-native weedy species, they provided a more "wild" aesthetic and required no artificial inputs.

13. Shutkin, *The Land That Could Be*.

14. Note that the Rouge River in Detroit is not the same Rouge River that flows through Scarborough, ON, mentioned in "The Principled Chapter."

15. K. G. Tidball et al., "Stewardship, Learning, and Memory in Disaster Resilience," *Environmental Education Research* 16, no. 5–6 (2010): 591–609.

16. G. A. Bonanno, "Loss, Trauma, and Human Resilience: How We Have Underestimated the Human Capacity to Thrive after Extremely Aversive Events," *American Psychologist* 59, no. 1 (2004): 20–28; A. S. Masten and J. Obradovic, "Disaster Preparation and Recovery: Lessons from Research on Resilience in Human Development," *Ecology and Society* 13, no. 1 (2008), http://www.ecologyand society.org/vol13/iss1/art9/.

17. H. Okvat and A. Zautra, "Sowing Seeds of Resilience: Community Gardening in a Post-Disaster Context," in *Greening in the Red Zone*, ed. K. G. Tidball and M. E. Krasny (New York: Springer, 2014), 73–90; "Community Gardening: A Parsimonious Path to Individual, Community, and Environmental Resilience," *American Journal of Community Psychology* 47 (2011): 374–387.

18. Story compiled from J. Dwyer, "New Daffodils for Garden That Outlived a Creator," *New York Times*, November 30, 2001, http://www.nytimes .com/2001/11/30/nyregion/a-nation-challenged-objects-new-daffodils-for-garden -that-outlived-a-creator.html; C. Hudson, "Community Hero: J. A. and Geraldine Reynolds," August 27, 2011, http://myhero.com/go/hero.asp?hero=Bruce_Gar den_Sept11.

## Steward Story: Monique Pilié

1. Story compiled from interviews and email correspondence with Monique Pilié over several years following Hurricane Katrina as part of Keith Tidball's dissertation research. See also Hike for KaTREEna, http://www.hikeforkatreena.com.

## 2  Love of Life, Love of Place

1. K. Helphand, *Defiant Gardens: Making Gardens in Wartime* (San Antonio, TX: Trinity University Press, 2006).

2. For a broader discussion of greening and defiance, see K. G. Tidball, "Peace Research and Greening in the Red Zone: Community-Based Ecological Restoration to Enhance Resilience and Transitions toward Peace," in *Expanding Peace Ecology: Peace, Security, Sustainability, Equity and Gender: Perspectives of IPRA's Ecology and Peace Commission*, ed. U. O. Spring, H. G. Brauch, and K. G. Tidball (Berlin: Springer-Verlag, 2014), 63–83.

3. E. Fromm, *The Heart of Man* (New York: Harper & Row, 1964). Italics in original.

4. Ibid., 41.

5. Ibid., 45.

6. Ibid., 47. According to Erich Fromm: "The opposite of the necrophilous orientation is the *biophilous*; its essence is love of life in contrast to love of death. Like necrophilia, biophilia is not constituted by a single trait, but represents a total orientation, an entire way of being. It is manifested in a person's bodily processes, in his emotions, in his thoughts, in his gestures; the biophilous orientation expresses itself in the whole man. The most elementary form of this orientation is expressed in the tendency of all living organisms to live" (45; italics in original).

7. E. O. Wilson, *Biophilia* (Cambridge, MA: Harvard University Press, 1984).

8. Ibid., 1.

9. S. R. Kellert and E. O. Wilson, *The Biophilia Hypothesis* (Washington, DC: Island Press, 1993).

10. Ibid.

11. E. Allen et al., "Against 'Sociobiology,'" *New York Review of Books* 182 (1975): 184–186, http://www.nybooks.com/articles/archives/1975/nov/13/against -sociobiology/; R. C. Lewontin, S. P. R. Rose, and L. J. Kamin, *Not in Our Genes: Biology, Ideology, and Human Nature* (New York: Pantheon, 1984); U. Segerstråle, *Defenders of the Truth: The Battle for Science in the Sociobiology Debate and Beyond* (Oxford: Oxford University Press, 2000).

12. P. Kitcher, *Vaulting Ambition: Sociobiology and the Quest for Human Nature* (Cambridge, MA: MIT Press, 1987); L. Leibowitz, "Humans and Other Animals: A Perspective on Perspectives," *Dialectical Anthropology* 9, nos. 1–4 (1985): 127–142.

13. This ability, or what some might refer to as an imperative, to cultivate biophilia in people who have little opportunity for contact with nature is closely aligned with calls to increase children's access to nature, and with research demonstrating how people seek out and benefit from such contact. See chapter 6 of this book; and L. Chawla, "Children's Engagement with the Natural World as a Ground for Healing," in *Greening in the Red Zone*, ed. K. G. Tidball and M. E. Krasny (New York: Springer, 2014), 111–124; N. Wells, "The Role of Nature in Children's Resilience: Cognitive and Social Processes," in *Greening in the Red Zone*, ed. Tidball and Krasny, 95–110; R. Louv, *Last Child in the Woods: Saving Our Children from Nature-Deficit Disorder* (New York: Algonquin Books, 2006).

14. J. A. Livingston, *The Fallacy of Wildlife Conservation* (Toronto: McClelland and Stewart, 1981), 114.

15. Ibid., 114.

16. A. Leopold, *A Sand County Almanac* (New York. Oxford University Press, 1949), 138–139.

17. A. Næss, "Self-Realization: An Ecological Approach to Being in the World," in *The Selected Works of Arne Næss*, ed. H. Glasser and A. Drengson (Dordrecht, The Netherlands: Springer, 2005), 517–518. Italics in original.

18. "Oneg Shabbat" refers to an informal Friday evening gathering of Jews in a synagogue or private home to express outwardly the happiness inherent in the Sabbath holiday. Literally, "Sabbath delight." Dictionary.com, "Oneg Shab-

bat," 2014, http://dictionary.reference.com/browse/oneg+shabbat; Encyclopedia Britannica, "Oneg Shabbat," February 19, 2014, http://www.britannica.com/EB-checked/topic/429113/Oneg-Shabbat.

19. S. S. Friedman, *Amcha: An Oral Testament of the Holocaust* (Washington, DC: University Press of America, 1979), 137.

20. K. G. Tidball, "Urgent Biophilia: Human-Nature Interactions in Red Zone Recovery and Resilience," in *Greening in the Red Zone*, ed. Tidball and Krasny: 53–73.

21. Tidball, "Urgent Biophilia"; M. E. Krasny et al., "Greening and Resilience in Military Communities," in *Greening in the Red Zone*, ed. Tidball and Krasny, 163–180.

22. Chawla, "Children's Engagement with the Natural World."

23. E. Lee and M. E. Krasny, "The Role of Social Learning for Social-Ecological Systems in Korean Village Groves Restoration," forthcoming in *Ecology and Society*; E. Lee, "Reconstructing Village Groves after a Typhoon in Korea," in *Greening in the Red Zone*, ed. Tidball and Krasny, 159–162.

24. L. Boukharaeva, "Six Ares of Land for the Resilience of Urban Families in Post-Soviet Russia," in *Greening in the Red Zone*, ed. Tidball and Krasny, 357–360.

25. Tidball and Krasny, *Greening in the Red Zone*.

26. Livingston, *Fallacy of Wildlife Conservation*, 100 (italics in original).

27. Y.-F. Tuan, *Topophilia* (Englewood Cliffs, NJ: Prentice-Hall, 1974). Topophilia is closely tied to place attachment and sense of place. Sense of place is a combination of place meaning, namely, the symbolic meanings that people ascribe to settings, and place attachment, namely, the bond between people and places or the degree to which a place is important to people. R. C. Stedman, "Toward a Social Psychology of Place: Predicting Behavior from Place-Based Cognitions, Attitude, and Identity."

28. Kellert and Wilson, *The Biophilia Hypothesis*.

29. According to Yi-Fu Tuan:

The word "topophilia" is a neologism, useful in that it can be defined broadly to include all of the human being's affective ties with the material environment. These differ greatly in intensity, subtlety, and mode of expression. The response to environment may be primarily aesthetic: it may then vary from the fleeting pleasure one gets from a view to the equally fleeting but far more intense sense of beauty that is suddenly revealed. The response may be tactile, a delight in the feel of air, water, earth. More permanent and less easy to express are feelings that one has toward a place because it is home, the locus of memories, and the means of gaining a livelihood. (*Topophilia*, 95)

For a discussion of expressions of biophilia and topophilia in disaster settings versus declining Rust Belt cities, see R. C. Stedman and M. Ingalls, "Topophilia, Biophilia and Greening in the Red Zone," in *Greening in the Red Zone*, ed. Tidball and Krasny, 129–144.

30. B. L. Amsden, R. C. Stedman, and L. E. Kruger, "The Creation and Maintenance of Sense of Place in a Tourism-Dependent Community," *Leisure Sciences* 33, no. 1 (2010): 32–51; L. C. Manzo and D. D. Perkins, "Finding Common Ground: The Importance of Place Attachment to Community Participation and Planning," *Journal of Planning Literature* 20, no. 4 (2006): 335–350.

31. Manzo and Perkins, "Finding Common Ground"; R. C. Stedman, "Toward a Social Psychology of Place: Predicting Behavior from Place-Based Cognitions, Attitude, and Identity," *Environment and Behavior* 34 (2002): 561–581.

32. K. Kindscher, "Gardening with Southeast Asian Refugees," 2009, http://kindscher.faculty.ku.edu/wp-content/uploads/2010/10/Kindscher-2009-Gardening-with-SE-Asian-Refugees-2.pdf.

33. Tuan, *Topophilia*. According to Tuan: "Topophilia is not the strongest of human emotions. When it is compelling we can be sure that the place or environment has become the carrier of emotionally charged events or perceived as a symbol" (95).

34. E. Gies, "Restoring Iraq's Garden of Eden," *New York Times,* April 17, 2013. http://www.nytimes.com/2013/04/18/world/middleeast/restoring-iraqs-garden-of-eden.html?pagewanted=all&_r=0.

35. K. G. Tidball and R. C. Stedman, "Positive Dependency and Virtuous Cycles: From Resource Dependence to Resilience in Urban Social-Ecological Systems," *Ecological Economics* 86 (2013): 292–299. Stedman and Ingalls, "Topophilia, Biophilia."

36. Through his work in post-Sandy New York, post-tornado Joplin, Missouri, and post-Katrina New Orleans, Keith Tidball and colleagues have defined mechanisms to explain greening in red zones in addition to urgent biophilia and restorative topophilia, including memorialization mechanisms, social-ecological symbols and social-ecological rituals, and discourses of defiance. K. G. Tidball, "Mechanisms of Resilience & Other 'Re-Words' in Urban Greening," April 24, 2013, http://www.thenatureofcities.com/2013/04/24/mechanisms-of-resilience-other-re-words-in-urban-greening; B. DuBois and K. G. Tidball, "Greening in Coastal New York Post-Sandy," Civic Ecology Lab White Paper Series, #CEL004 (Ithaca, NY: Cornell Civic Ecology Lab, Cornell University, 2013). http://civeco.files.wordpress.com/2013/09/2013-dubois.pdf.

37. Leopold, *Sand County Almanac*; K. G. Tidball, "Greening in the Red Zone: Valuing Community-Based Ecological Restoration in Human Vulnerability Contexts," Ph.D. (Ithaca, NY: Cornell University, 2012).

38. P. Cafaro, "Thoreau, Leopold, and Carson: Toward an Environmental Virtue Ethics," *Environmental Ethics* 22, no. 4 (2001): 3–17.

39. Philip Cafaro uses the term "environmental virtue ethics" to describe living well in nature, based on the works of Henry David Thoreau, Aldo Leopold, and Rachel Carson. Such an ethic "makes room in our lives for the rest of creation." On top of arguments for preserving nature to satisfy our materialistic needs (for example, as a source of new medicines or lumber) or for nature's intrinsic value,

an environmental virtue ethics suggests that we must preserve nature "to preserve human possibilities and help us become better people." Ibid.

Similarly, the Norwegian philosopher Arne Næss, reflecting his notion of the "ecological self," believed that asking people to give something up is a "treacherous basis for conservation" but that conservation based on "the love of a widened and deepened self" will better serve the conservation cause. Civic ecology practices provide opportunities to "live well in nature"—to develop a sense of our selves in nature—and thus provide a promising path toward a conservation ethic. A. Næss, "Self-Realization: An Ecological Approach to Being in the World," in *The Selected Works of Arne Næss*, ed. H. Glasser and A. Drengson (Dordrecht, The Netherlands: Springer, 2005), 515–530.

40. Leopold, *Sand County Almanac*.

41. Næss, "Self-Realization," 519.

42. Ibid.

43. Fromm, *Heart of Man*.

44. Ibid., 47.

45. Cafaro, "Thoreau, Leopold, and Carson."

46. Helphand, *Defiant Gardens*.

## Steward Story: Nilka Martell

1. Story compiled from email and phone conversations with Martell in February 2014, and the following sources. H. Sheffield, "How One Unemployed Bronxite Transformed Her Block," October 20, 2011, http://hazelsheffield.com/tag/nilka -martell; J. Evelly, "Bronx Group Turns Parking Space into Temporary Public 'Park,'" *DNAinfo New York*, September 26, 2012, http://www.dnainfo.com/new -york/20120926/soundview/bronx-group-turns-parking-space-into-temporary -public-park; H. Sheffield, "Turning Unemployment into Time Well Spent," *The Bronx Ink*, November 3, 2011, http://bronxink.org/2011/11/03/19123-turning -unemployment-into-time-well-spent/; P. Wall, "Parkchester Block Beautification Group Branches out across the Bronx," September 4, 2013, http://www.dnainfo .com/new-york/20130904/parkchester/parkchester-block-beautification-group -branches-out-across-bronx.

2. NYC Service, http://www.nycservice.org.

3. M. Powell, "Bronx River Sankofa: Eco-Cultural History in New York City!," April 8, 2004, http://bronxriversankofa.wordpress.com.

## The First Interlude

1. In the *Oxford Dictionary*, "bricolage" is defined as construction or creation from a diverse range of available things. Oxford Dictionary, "Bricolage," n.d., http://www.oxforddictionaries.com/definition/english/bricolage.

## 3    Creating Community, Creating Connections

1. G. Kroth, "Guatemala's Garbage Dump Education System," *CounterPunch*, August 5, 2009, http://www.counterpunch.org/2009/08/05/guatemala-s-garbage -dump-education-system/; M. Nathe, "Guatemala City's Dirtiest Secret," *Endeavors*, January 14, 2008, http://endeavors.unc.edu/win2008/guatemala.php.

2. D. Winterbottom, "Developing a Safe, Nurturing and Therapeutic Environment for the Families of the Garbage Pickers in Guatemala and for Disabled Children in Bosnia and Herzegovina," in *Greening in the Red Zone*, ed. K. G. Tidball and M. E. Krasny (New York: Springer, 2014), 397–410.

3. N. Lemann, *The Promised Land: The Great Black Migration and How It Changed America* (New York: Random House, 1991).

4. Ibid., 107.

5. You can view a time-lapse video of the demolition of the Robert Taylor Homes at http://www.youtube.com/watch?v=z5ZHYcStqac.

6. One's sense of community may be linked to one's sense of place. B. L. Amsden, R. C. Stedman, and L. E. Kruger, "The Creation and Maintenance of Sense of Place in a Tourism-Dependent Community," *Leisure Sciences* 33, no. 1 (2010): 32–51. Sense of place also bears similarities to Tuan's concept of topophilia. See chapter 2 and R. C. Stedman, "Toward a Social Psychology of Place: Predicting Behavior from Place-Based Cognitions, Attitude, and Identity," *Environment and Behavior* 34 (2002): 561–581; Y.-F. Tuan, *Topophilia* (Englewood Cliffs, NJ: Prentice-Hall, 1974).

7. Information about Nature Cleaners was collected by Zahra Golshani during a visit to Iran in summer 2013.

8. L. J. Hanifan, "The Rural School Community Center," *Annals of the American Academy of Political and Social Science* 67 (1916): 130–138.

9. Refer to G. Hardin, "The Tragedy of the Commons," *Science* 162, no. 3859 (1968): 1243–1248.

10. Hanifan, "Rural School Community Center," 130.

11. Ibid., 130–131. Hanifan's notions about individual and community benefits of social capital foreshadow a debate among social scientists that still today frames much of the discussion: Is social capital an attribute of individuals that they "cash in on" for their own benefit (like using connections to get a job), or of communities, cities, regions, and countries that predicts outcomes related to the health of the society (such as better schools)? Social capital also can be an attribute of small groups like families or people sharing the management of a resource such as a village forest. See D. Castiglione, J. W. van Deth, and G. Wolleb, eds., *The Handbook of Social Capital* (Oxford: Oxford University Press, 2008).

12. Hanifan, "Rural School Community Center."

13. Ibid., 130.

14. Saw Mill River Coalition, "Daylighting the Saw Mill River in Yonkers," n.d., http://www.sawmillrivercoalition.org/whats-happening/daylighting-the-saw-mill -river-in-yonkers.

15. R. B. Primack, H. Kobori, and S. Mori, "Dragonfly Pond Restoration Promotes Conservation Awareness in Japan," *Conservation Biology* 14, no. 5 (2000): 1153–1554.

16. Jerusalem Municipality, "The Gazelle Valley Urban Nature Park," n.d., http://greenpilgrimjerusalem.org/symposium2013/wp-content/uploads/sites/3/2013/02/Gazelle-Valley-Intro.pdf.

17. T. K. Ahn and E. Ostrom, eds., *Foundations of Social Capital* (Chelthenham, UK: Edward Elgar Publishing, 2003); Ahn and Ostrom, "Social Capital and Collective Action," in *Handbook of Social Capital*, 70–100. According to Ostrom, "when individuals are well informed about the problem they face and about who else is involved, and can build settings where trust and reciprocity can emerge, grow, and be sustained over time, costly and positive actions are frequently taken without waiting for an external authority to impose rules, monitor compliance, and assess penalties." E. Ostrom, "Polycentric Systems for Coping with Collective Action and Global Environmental Change," *Global Environmental Change* 20 (2010): 555. In addition to social capital, the following conditions predict when management of a common property resource is likely to be sustainable: the resource and its use can be monitored; rates of change in the resource, its users, technologies, and socioeconomic context are moderate; outsiders can be excluded from using the resources; and users support monitoring and enforcing the rules. T. Dietz, E. Ostrom, and P. C. Stern, "The Struggle to Govern the Commons," *Science* 302, no. 5652 (December 12, 2003): 1907–1912.

18. R. B. Putnam, *Bowling Alone: The Collapse and Revival of American Community*; R. B. Putnam, "Bowling Alone: America's Declining Social Capital," *Journal of Democracy* 6, no. 1 (1995): 65–78.

19. Putnam, "Bowling Alone: America's Declining Social Capital."

20. Saguaro Seminar, "Social Capital Community Benchmark Survey: Executive Summary" (Cambridge, MA: Harvard University, n.d.), http://www.hks.harvard.edu/saguaro/communitysurvey/docs/exec_summ.pdf.

21. Social capital manifests itself in features of social organization, including networks, norms, social trust, and civic engagement, which pervasively influence our public life as well as our private prospects. Putnam, "Bowling Alone: America's Declining Social Capital"; A. Portes, "Social Capital: Its Origins and Applications in Modern Sociology," *Annual Review of Sociology* 24 (1998); A. Portes and P. Landolt, "Social Capital: Promise and Pitfalls of Its Role in Development," *Journal of Latin American Studies* 32, no. 2 (2000).

Social capital has been correlated with, among other positive community attributes, lower crime rates, higher performance of civic institutions, and longevity, and lower dropout rates. See K. Lochner, I. Kawachi, and B. P. Kennedy, "Social Capital: A Guide to Its Measurement," *Health & Place* 5 (1999); J. S. Coleman, "Social Capital in the Creation of Human Capital," *The American Journal of Sociology* 94 (Supplement) (1988): S95–S120.

22. Putnam, "Bowling Alone: America's Declining Social Capital," 67. Italics added.

23. Saguaro Seminar, "Social Capital Community Benchmark Survey."

24. How are all eleven of Putnam's social capital domains expressed in civic ecology practices? By pulling weeds, preparing soil, and replanting native species in Seattle's city parks, civic ecology stewards not only become involved in a friends of the park civic association (civic leadership [4], associational involvement [5]), they also learn to trust (social trust [1]) the person who hands them the shovel or helps them to navigate the wheelbarrow down a precipitous, rutted pathway.

Time spent under an oak tree on a sultry day in a community garden in Houston enables gardeners and visitors to trade stories (informal socializing [8], diversity of friendship network [3], interracial trust [2]), and volunteer gardeners often bring a share of their bounty of vegetables and herbs to elderly housebound residents living nearby (giving and volunteering [9]).

Friends of the Rouge Watershed, a civic ecology organization outside the city of Toronto, sponsors interfaith Muslim-Jewish tree planting days, and advocates for formal recognition of the watershed they are restoring as the world's largest urban park (faith-based social engagement [6], equality of civic engagement across the community [7], conventional and protest politics participation [10, 11]).

25. A. de Tocqueville, *Democracy in America* (Cambridge, MA Harvard University Press, 1863); D. Castiglione, "Social Capital as a Research Programme," in *Handbook of Social Capital.*

26. C. Sirianni and L. A. Friedland, *The Civic Renewal Movement: Community Building and Democracy in the United States* (Dayton, OH: Charles F. Kettering Foundation, 2005); R. J. Sampson et al., "Civil Society Reconsidered: The Durable Nature and Community Structure of Collective Civic Action," *American Journal of Sociology* 111, no. 3 (2005): 673–714.

27. Sirianni and Friedland, *Civic Renewal Movement*, 122.

28. Sampson et al., "Civil Society Reconsidered," 674.

29. S. Ansari, "Social Capital and Collective Efficacy: Resource and Operating Tools of Community Social Control," *Journal of Theoretical and Philosophical Criminology* 5, no. 2 (2013): 75–94. Ansari details the commonalities and differences between social capital and collective efficacy. Social capital emphasizes social networks, participation, and trust and reciprocity. Collective efficacy focuses on expectations for social control and willingness to intervene for the public good and thus, switches the emphasis from a "capital" or entity that people use, to how a community or neighborhood takes it upon itself to fight against crime and disorder. "The essence of collective efficacy is that residents will have willingness to intervene for the common good if they share mutual trust and solidarity. It means if residents, living in one neighborhood, trust one another and are ready to help and reciprocate, they will be willing to intervene in case of a danger or any unpleasant and emergency situation." Ibid., 82.

30. C. Lewis, "Public Housing Gardens: Landscapes for the Soul," in *Landscape for Living* (Washington, DC: USDA Yearbook of Agriculture, 1972); C. Lewis and S. Sturgill, "Comment: Healing in the Urban Environment: A Person/Plant Viewpoint," *Journal of the American Planning Association* 45, no. 3 (1979): 330–338.

31. Lewis, "Public Housing Gardens."

32. Lewis and Sturgill, "Comment," 332.

33. D. Harper, "Online Etymology Dictionary," 2014, http://www.etymonline.com.

34. D. Hurley, "Scientist at Work—Felton Earls; On Crime as Science (a Neighbor at a Time)," *New York Times*, January 6, 2004, http://www.nytimes.com/2004/01/06/science/scientist-at-work-felton-earls-on-crime-as-science-a-neighbor-at-a-time.html; R. J. Sampson, S. W. Raudenbush, and F. Earls, "Neighborhoods and Violent Crime: A Multilevel Study of Collective Efficacy," *Science* 277, no. 5328 (1997): 918–924.

35. G. L. Kelling and J. Q. Wilson, "Broken Windows," *The Atlantic*, March 1982, 29–38. According to Kelling and Wilson, "At the community level, disorder and crime are usually inextricably linked, in a kind of developmental sequence. Social psychologists and police officers tend to agree that if a window in a building is broken and is left unrepaired, all the rest of the windows will soon be broken. . . . [O]ne unrepaired broken window is a signal that no one cares, and so breaking more windows costs nothing" (2–3).

36. A controlled study conducted in the city of Groningen, The Netherlands, provides strong evidence for the "broken windows" hypothesis. Researchers observed "incivil" (e.g., littering) and criminal (stealing) behaviors of people in the presence and absence of graffiti, litter, illegally parked bicycles, and abandoned shopping carts, and found that, "when people observe that others violated a certain social norm or legitimate rule, they are more likely to violate other norms or rules, which causes disorder to spread." K. Keizer, S. Lindenberg, and L. Steg, "The Spreading of Disorder," *Science* 322 (2008): 1681–1684.

37. Sampson, Raudenbush, and Earls, "Neighborhoods and Violent Crime."

38. Hurley, "Scientist at Work."

39. Ibid.

40. L. Saldivar and M. E. Krasny, "The Role of NYC Latino Community Gardens in Community Development, Open Space, and Civic Agriculture," *Agriculture and Human Values* 21 (2004): 399–412. See also the steward stories of Nilka Martell and Andre Rivera from the Bronx.

41. Sampson, Raudenbush, and Earls, "Neighborhoods and Violent Crime."

42. Work on management of common pool (or shared) resources focuses on resources that communities depend on for their economic livelihoods, whereas the focus of civic ecology practices is on management of resources people depend on largely for emotional sustenance, including the need to connect with nature, place, and people. See K. G. Tidball and R. C. Stedman, "Positive Dependency and Virtuous Cycles: From Resource Dependence to Resilience in Urban Social-Ecological Systems," *Ecological Economics* 86 (2013): 292–299.

43. A. Leopold, *Sand County Almanac* (New York: Oxford University Press, 1949), 240.

44. N. Noddings, "On Community," *Educational Theory* 46, no. 3 (1996): 245–267.

45. Putnam, *Bowling Alone: The Collapse and Revival of American Community*.

46. Sampson, Raudenbush, and Earls, "Neighborhoods and Violent Crime."

## Steward Story: Kazem Nadjariun

1. Story compiled from Kazem Nadjariun's answers to questions sent via email, translated from Farsi by Somayeh Khiyabani with help from Zahra Golshani. Golshani established connections with Nadjariun first through Facebook and later while on a visit to Iran in the summer of 2013.

## 4 Oyster Spat and Live Oaks—Memories

1. J. Mitchell, *The Bottom of the Harbor* (New York: Random House, Inc., 1951), 75.

2. Ibid.

3. According to Mitchell, "A few elderly men who once were bedders are still living in the old Staten Island oyster ports, and many sons and grandsons of bedders. They have a proprietary feeling about harbor oysters, and every so often, in cold weather, despite the laws, some of them go out to the old, ruined beds and poach a mess. They know what they are doing; they watch the temperature of the water to make sure the oysters are "sleeping," or hibernating, before they eat any" (ibid., 55–56).

4. Ibid., 54.

5. M. Kurlansky, *The Big Oyster: History on the Half Shell* (New York: Ballantine Books, 2006).

6. L. Hansen, "'History on the 'Half Shell' in 'Big Oyster,'" *National Public Radio*, April 6, 2006, http://www.npr.org/templates/story/story.php?storyId=5332982.

7. The term "ecological memories" is also used to refer to remnant populations of organisms. F. Berkes and C. Folke, "Back to the Future: Ecosystem Dynamics and Local Knowledge," in *Panarchy: Understanding Transformations in Human and Natural Systems*, ed. L. H. Gunderson and C. S. Holling (Washington, DC: Island Press, 2002), 121–146; M. Gadgil, N. S. Hemam, and B. M. Reddy, "People, Refugia and Resilience," in *Linking Social and Ecological Systems: Management Practices and Social Mechanisms for Building Resilience*, ed. F. Berkes and C. Folke (Cambridge, UK: Cambridge University Press, 1998), 30–47; M. Gadgil et al., "Exploring the Role of Local Ecological Knowledge in Ecosystem Management: Three Case Studies," in *Navigating Social-Ecological Systems: Building Resilience for Complexity and Change*, ed. F. Berkes, J. Colding, and C. Folke (Cambridge, UK: Cambridge University Press, 2003), 189–209.

8. R. J. McIntosh, J. A. Tainter, and S. K. McIntosh, "Climate, History, and Human Action," in *The Way the Wind Blows: Climate, History, and Human Action*, ed. R. J. McIntosh, J. A. Tainter, and S. K. McIntosh (New York: Columbia University Press, 2000), 1–42.

9. Kurlansky, *The Big Oyster*; Mitchell, *The Bottom of the Harbor*. In addition to story telling, social memories may be encoded in symbols, legends and beliefs, harvest ceremonies and other rituals, and in actual hunting, gardening, tree planting, and other resource management practices. The live oaks that were symbolic of New Orleans played a role in transmitting memories of the role of green spaces in bringing a community together—one that was to be revisited after the Hurricane Katrina disaster. K. G. Tidball, "Greening in the Red Zone: Valuing Community-Based Ecological Restoration in Human Vulnerability Contexts," Ph.D. diss. (Ithaca, NY: Cornell University, 2012).

10. S. Barthel et al., "History and Local Management of a Biodiversity-Rich, Urban Cultural Landscape," *Ecology and Society* 10, no. 2 (2005), http://www.ecology andsociety.org/vol10/iss2/art10/; S. Barthel, C. Folke, and J. Colding, "Social-Ecological Memory in Urban Gardens: Retaining the Capacity for Management of Ecosystem Services," *Global Environmental Change* 20 (2010): 255–265.

11. Kurlansky, *The Big Oyster*.

12. Gadgil, Hemam, and Reddy, "People, Refugia and Resilience."

13. Social-ecological memories are the "knowledge, experience and practice about how to manage a local ecosystem and its services (that) is retained in a community, and modified, revived and transmitted through time" Barthel, Folke, and Colding, "Social-Ecological Memory in Urban Gardens."

14. F. Berkes, "Knowledge, Learning and the Resilience of Social-Ecological Systems," in *Knowledge for the Development of Adaptive Co-Management* (Oaxaca, MX: Tenth Biennial Conference of the International Association for the Study of Common Property, 2004), http://dlc.dlib.indiana.edu/dlc/bitstream/handle /10535/2385/berkes_knowledge_040511_paper299a.pdf?sequence=1.

15. McIntosh, Tainter, and McIntosh, "Climate, History, and Human Action," 24–25.

16. K. G. Tidball et al., "Stewardship, Learning, and Memory in Disaster Resilience," *Environmental Education Research* 16, nos. 5–6 (2010): 591–609.

17. D. Rogers, "Planners Push to Tear out Elevated I-10 over Claiborne," *Times-Picayune*, July 11, 2009, http://floor8.info/nola/20090700/090711_proposal_to_ tear_down_i_10_over_claiborne.html.

18. Quote is from a Tremé community leader and tree planter, interviewed by Keith Tidball, January 19, 2009. K. G. Tidball, "Greening in the Red Zone: Valuing Community-Based Ecological Restoration in Human Vulnerability Contexts."

19. K. G. Tidball et al., "Stewardship, Learning, and Memory in Disaster Resilience."

20. Tidball, "Greening in the Red Zone: Valuing Community-Based Ecological Restoration in Human Vulnerability Contexts."

21. Tidball has described in detail how social-ecological memories become embedded in social-ecological symbols (e.g., trees) and rituals (tree planting) in post-Katrina New Orleans. K. G. Tidball, "Trees and Rebirth: Social-Ecological Symbols and Rituals in the Resilience of Post-Katrina New Orleans, " in *Greening in the Red Zone*, ed. K. G. Tidball and M. E. Krasny, 257–296 (New York: Springer,

2014); K. G. Tidball, "Seeing the Forest for the Trees: Hybridity and Social-Ecological Symbols, Rituals and Resilience in Post-Disaster Contexts," forthcoming in *Ecology and Society* (special issue on Exploring Social-Ecological Resilience through the Lens of the Social Sciences: Contributions, Critical Reflections and Constructive Debate).

## 5   Ecosystem Services

1. S. Díaz et al., "Linking Functional Diversity and Social Actor Strategies in a Framework for Interdisciplinary Analysis of Nature's Benefits to Society," *Proceedings of the National Academy of Sciences* 108, no. 3 (2011): 896.

2. E. O. Wilson, *Biophilia* (Cambridge, MA: Harvard University Press, 1984).

3. M. J. Peterson et al., "Obscuring Ecosystem Function with Application of the Ecosystem Services Concept," *Conservation Biology* 24, no. 1 (2010): 113–119.

4. Ibid. E. Gómez-Baggethun et al., "The History of Ecosystem Services in Economic Theory and Practice: From Early Notions to Markets and Payment Schemes," *Ecological Economics* 69 (2010): 1209–1218.

5. We don't mean to imply that education causes behavior change. For a review of research demonstrating otherwise, see T. A. Heberlein, *Navigating Environmental Values* (Oxford: Oxford University Press, 2012).

6. R. Costanza et al., "The Value of the World's Ecosystem Services and Natural Capital," *Nature* 387 (1997): 253–260.

7. Gómez-Baggethun et al., "The History of Ecosystem Services in Economic Theory and Practice"; Peterson et al., "Obscuring Ecosystem Function"; S. Díaz et al., "Linking Functional Diversity and Social Actor Strategies in a Framework for Interdisciplinary Analysis of Nature's Benefits to Society." Despite the fact that people are attracted by and even worship plants, birds, mammals, and insects, when viewing biological diversity from an ecosystem services perspective, the individual species is not so important. Rather its function in the ecosystem—for example, decomposition, water filtration, pollination, food production, carbon sequestration, or recreation—is what is important. Scientists have coined the term "functional diversity" to emphasize these ecosystem roles or functions of organisms.

8. Paul Stern differentiates between social-altruistic, biospheric, and egocentric values. Valuing nature because of its economic contributions to humans would fall under his social-altruistic or egocentric category, whereas valuing nature for its intrinsic worth without regard to human needs is consistent with biospheric values. P. C. Stern, "The Value Basis of Environmental Concern," *Journal of Social Issues* 50, no. 3 (1994): 65–84.

9. Gómez-Baggethun et al., "The History of Ecosystem Services in Economic Theory and Practice."

10. K. M. A. Chan et al., "Where Are Cultural and Social in Ecosystem Services? A Framework for Constructive Engagement," *BioScience* 62, no. 8 (2012): 744–756; K. M. A. Chan, T. Satterfield, and J. Goldstein, "Rethinking Ecosystem Ser-

vices to Better Address and Navigate Cultural Values," *Ecological Economics* 74 (2012): 8–18; M. Kumar and P. Kumar, "Valuation of the Ecosystem Services: A Psycho-Cultural Perspective," *Ecological Economics* 64, no. 4 (2008): 808–819; M. A. Wilson and R. B. Howarth, "Discourse Based Valuation of Ecosystem Services: Establishing Fair Outcomes through Group Deliberation," *Ecological Economics* 41 (2002): 431–443.

11. Arne Næss introduced the term "ecological self" to describe how humans identify with and have a deep empathy for other living beings. According to Næss: "Traditionally the *maturity of self* has been considered to develop through three stages, from ego to social self, comprising the ego, and from there to the metaphysical self, comprising the social self. But Nature is then largely left out in the conception of this process. Our home, our immediate environment, where we belong as children, and the identification with human living beings, are largely ignored. I therefore tentatively introduce, perhaps for the first time ever, a concept of *ecological self*. We may be said to be in, of and for Nature from our very beginning" (A. Næss, "Self-Realization: An Ecological Approach to Being in the World," in *The Selected Works of Arne Næss*, ed. Harold Glasser and Alan Drengson [Dordrecht, The Netherlands: Springer, 2005], 516).

12. M. E. Krasny et al., "Civic Ecology Practices: Participatory Approaches to Generating and Measuring Ecosystem Services in Cities," *Ecosystems Services* 7 (March 2014): 177–186; B. Reyers et al., "Getting the Measure of Ecosystem Services: A Social-Ecological Approach," *Frontiers in Ecology and Environment* 11, no. 5 (2013): 268–273.

13. MEA, *Millennium Ecosystem Assessment: Ecosystems and Human Well-Being: Synthesis* (Washington, DC: Island Press, 2005).

14. Ibid.

15. T. Elmqvist et al., *Urbanization, Biodiversity and Ecosystem Services: Challenges and Opportunities* (New York: Springer, 2013).

16. A. Mancabelli, "Symbolic Meaning and Emotion in Japanese Ethnozoology," paper presented at the Annual Ethnobiological Conference, Fairbanks, AK, May 2002; R. B. Primack, H. Kobori, and S. Mori, "Dragonfly Pond Restoration Promotes Conservation Awareness in Japan," *Conservation Biology* 14, no. 5 (2000): 1153–1154.

17. The human-shaped landscapes also had a greater diversity of plants and wildlife than did forested areas not bearing such a strong imprint by humans. See H. Kobori and R. B. Primack, "Participatory Conservation Approaches for *Satoyama*, the Traditional Forest and Agricultural Landscape of Japan," *Ambio* 32, no. 4 (2003): 307–311.

18. Ibid.

19. Ibid.

20. Primack, Kobori, and Mori, "Dragonfly Pond Restoration"; H. Kobori, "Current Trends in Conservation Education in Japan," *Biological Conservation* 142 (2009): 1950–1957.

21. Rocking the Boat (RTB), n.d., http://www.rockingtheboat.org/.

22. Krasny et al., "Civic Ecology Practices."

23. E. Andersson, S. Barthel, and K. Ahrne, "Measuring Social-Ecological Dynamics Behind the Generation of Ecosystem Services," *Ecological Applications* 17, no. 5 (2007): 1267–1278.

24. E. Strauss, "Urban Pollinators and Community Gardens," *Cities and the Environment* 2, no. 1 (2009), http://digitalcommons.lmu.edu/cate/vol2/iss1/1/.

25. L. J. Lawson, *City Bountiful: A Century of Community Gardening in America* (Berkeley: University of California Press, 2005); L. Saldivar and M. E. Krasny, "The Role of NYC Latino Community Gardens in Community Development, Open Space, and Civic Agriculture," *Agriculture and Human Values* 21 (2004): 399–412.

26. M. E. Krasny and K. G. Tidball, "Community Gardens as Contexts for Science, Stewardship, and Civic Action Learning," *Cities and the Environment* 2, no. 1 (2009), http://digitalcommons.lmu.edu/cgi/viewcontent.cgi?article=1037&context=cate.

27. NY/NJ Baykeeper, n.d., http://www.nynjbaykeeper.org.

28. D. E. Pataki et al., "Coupling Biogeochemical Cycles in Urban Environments: Ecosystem Services, Green Solutions, and Misconceptions," *Frontiers in Ecology and Environment* 9, no. 1 (2011): 27–36; K. L. Wolf et al., "Environmental Stewardship Footprint Research: Linking Human Agency and Ecosystem Health in the Puget Sound Region," *Urban Ecosystems* (2011), http://www.naturewithin.info/CivicEco/UrbanEcosystems.Stewardship.Apr2011online.pdf; PlaNYC, "MillionTreesNYC."

29. PlaNYC, "MillionTreesNYC."

30. K. L. Wolf et al., "Environmental Stewardship Footprint Research"; P. Silva and M. E. Krasny, "Parsing Participation: Models of Engagement for Outcomes Monitoring in Urban Stewardship," *Local Environment: The International Journal of Justice and Sustainability*, 2014, http://dx.doi.org/10.1080/13549839.2014.929094.

31. N. Tyack, "A Trash Biography" (Los Angeles: Friends of the Los Angeles River, 2011), http://folar.org/wp-contentuploads/2011/11/trashsortreport.pdf.

32. EarthCorps, "Local Restoration, Global Leadership: EarthCorps," n.d., http://www.earthcorps.org/index.php.

33. Compost for Brooklyn, n.d., http://compostforbrooklyn.org/about.

34. Even though Feedback Farms focuses on urban agriculture, its mission incorporates a broader commitment to environment and community, consistent with civic ecology practice. Feedback Farms, n.d., http://www.feedbackfarms.com.

35. Silva and Krasny, "Parsing Participation"; P. Silva and M. E. Krasny, "Outcomes Monitoring within Civic Ecology Practices in the NYC Region" (Ithaca, NY: Civic Ecology Lab, Cornell University, 2013), http://civeco.files.wordpress.com/2013/09/silva-2013.pdf.

36. U.S. Forest Service, "I-Tree," n.d., http://www.itreetools.org/.

37. Gowanus Canal Conservancy, n.d., http://www.gowanuscanalconservancy .org/ee.

38. "Gowanus Canal Conservancy Blog," June 27, 2014, http://gowanus canalconservancy.wordpress.com.

39. S. Sassen and N. Dotan, "Delegating, Not Returning, to the Biosphere: How to Use the Multi-Scalar and Ecological Properties of Cities," *Global Environmental Change* 21 (2011): 823–834; K. Tzoulas et al., "Promoting Ecosystem and Human Health in Urban Areas Using Green Infrastructure: A Literature Review," *Landscape and Urban Planning* 81 (2007): 167–178; TEEB, "TEEB—the Economics of Ecosystems and Biodiversity for Local and Regional Policy Makers," 2010, http://www.teebweb.org/publication/teeb-for-local-and-regional-policy-makers -2/; P. Bolund and S. Hunhammar, "Ecosystem Services in Urban Areas," *Ecological Economics* 29 (1999): 293–301; Pataki et al., "Coupling Biogeochemical Cycles in Urban Environments"; Elmqvist et al., *Urbanization, Biodiversity and Ecosystem Services*.

40. E. Lee, "Reconstructing Village Groves after a Typhoon in Korea," in *Greening in the Red Zone*, ed. K. G. Tidball and M. E. Krasny (New York: Springer, 2014), 159–162.

41. Díaz et al., "Linking Functional Diversity."

42. Krasny et al., "Civic Ecology Practices"; Reyers et al., "Getting the Measure of Ecosystem Services."

43. For another type of circularity or feedback between civic ecology practices, ecosystem services, and learning, see chapter 7, and K. G. Tidball and M. E. Krasny, "Toward an Ecology of Environmental Education and Learning," *Ecosphere* 2, no. 2 (2011), http://www.esajournals.org/doi/pdf/10.1890/ES10-00153.1.

## Steward Story: Marian Przybilla and Helga Garduhn

1. Profile largely compiled from V. J. Becker, "Regreening Instead of Guarding," *Neues Deutschland Sozialistische Tageszeitung*, trans. Bernd Blossey, August 14, 2010, http://www.neues-deutschland.de/artikel/177381.begruenen-statt -bewachen.html; and from H. W. White, "Bergfelde: The Teacher and Conservationist Helga Garduhn Is Now 75 Years Old," *Markische Online Zeitung*, August 9, 2011.

2. History.com, "Berlin Wall," n.d., http://www.history.com/topics/cold-war/ berlin-wall.

3. M. Cramer, "The Berlin Wall Trail: A Cycling and Hiking Route on the Traces of Berlin's East-West Division during the Cold War," in *Greening in the Red Zone*, ed. K. G. Tidball and M. E. Krasny (New York: Springer, 2014), 445–449.

## 6   Stewardship, Health, and Well-Being

1. Quote from an interview conducted February 2, 2014, with Saody Ouch, Cambodian refugee and community organizer with the Lowell Alliance. Ouch was speaking about Cambodian, Laotian, and Burmese refugees at the Franklin Court Community Garden, in Lowell, Massachusetts.

2. This passage goes on to describe how Mandela's work in the garden—evidence of self-efficacy and empowerment—extended to other prisoners. "Much of his correspondence with the authorities reveals the adroit ways in which he extended his influence from the tiny patch of soil in the corner of a courtyard to the entire prison, the other inmates and the warders, his many connections beyond the prison walls, the apartheid state, and ultimately the world." Interestingly, when a fellow prisoner wrote, "'Mr. Mandela is our chief gardener,'" the authorities "recognised the power in that seemingly innocuous line . . . as it was censored out of the letter." From Nelson Mandela Foundation, *A Prisoner in the Garden* (New York: Viking Studio, Penguin Group, 2006). 205.

3. G. H. Donovan et al., "The Relationship between Trees and Human Health: Evidence from the Spread of the Emerald Ash Borer," *American Journal of Preventive Medicine* 44, no. 2 (2013): 139–145.

4. C. C. Branas et al., "A Difference-in-Differences Analysis of Health, Safety, and Greening Vacant Urban Space," *American Journal of Epidemiology* 174, no. 11 (2011): 1296–1306.

5. C. Geisler, "Green Zones from Above and Below: A Retrospective Cautionary Tale," in *Greening in the Red Zone*, ed. K. G. Tidball and M. E. Krasny (New York: Springer, 2014), 203–215.

6. This information comes from several sources cited in T. S. Eisenman, "Frederick Law Olmsted, Green Infrastructure, and the Evolving City, " *Planning History* 12 (2013): 287–311. The sources include J. Martin, "Genius of Place: The Life of Frederick Law Olmsted" (Cambridge, MA: Da Capo Press, 2011); and F. L. Olmsted and C. Vaux, "A Review of Recent Changes, and Changes which have been Projected, in the Plans of the Central Park, by the Landscape Architects, 1872," in *F. L. Olmsted, Landscape Architect, 1822–1903*, vol. 2: *Central Park as a Work of Art and as a Great Municipal Enterprise 1853–1895*, 3rd ed., ed. Frederick Law Olmsted Jr. and Theodora Kimball (Bronx, NY: Benjamin Blom, Inc., 1970), 249. The quotation comes from the latter source.

7. Measures of recovery in the Ulrich (1984) study included: number of days of hospitalization; number and strength of analgesic doses for anxiety each day (including tranquilizers and barbiturates); minor complications, such as persistent headache and nausea; and nurses' notes relating to a patient's condition or course of recovery. See R. S. Ulrich, "View through a Window May Influence Recovery from Surgery," *Science* 224 (1984): 420–421.

8. N. Wells and K. A. Rollings, "The Natural Environment in Residential Settings: Influences on Human Health and Function," in *The Oxford Handbook of Environmental and Conservation Psychology*, ed. S. Clayton (Oxford: Oxford Univer-

sity Press, 2012), 509–523; R. Kaplan and S. Kaplan, *The Experience of Nature: A Psychological Perspective* (Cambridge: Cambridge University Press, 1989).

9. Solastalgia, defined as "the distress that is produced by environmental change impacting on people while they are directly connected to their home environment," has been described in drought and mining contexts in Australia. Relevant to our discussions of stewardship, self-efficacy, and empowerment, solastalgia can lead to a "sense of powerlessness or lack of control over the unfolding change process." G. Albrecht et al., "Solastalgia: The Distress Caused by Environmental Change," *Australasian Psychiatry* 15 (2007): S95–S98.

10. T. Hartig, M. Mang, and G. W. Evans, "Restorative Effects of Natural Environment Experiences," *Environment and Behavior* 23, no. 1 (1991): 3–26; R. Kaplan and S. Kaplan, *The Experience of Nature: A Psychological Perspective* (Cambridge, UK: Cambridge University Press, 1989); P. D. Relf, "The Therapeutic Values of Plants," *Pediatric Rehabilitation* 8, no. 3 (2005): 235–237.

11. A. Faber Taylor and F. E. Kuo, "Children with Attention Deficits Concentrate Better after Walk in the Park," *Journal of Attention Disorders* 12 (2009): 402–409; A. Faber Taylor et al., "Growing up in the Inner City: Green Spaces as Places to Grow," *Environment and Behavior* 30, no. 1 (1998): 3–27; A. Faber Taylor, F. E. Kuo, and W. C. Sullivan, "Coping with ADD: The Surprising Connection to Green Play Settings," *Environment and Behavior* 33, no. 1 (2001): 54–77; N. Wells, "The Role of Nature in Children's Resilience: Cognitive and Social Processes," in *Greening in the Red Zone*, ed. K. G. Tidball and M. E. Krasny, 95–110 (New York: Springer, 2014); N. Wells, "At Home with Nature—Effects of Greenness on Children's Cognitive Functioning," *Environment and Behavior* 32 (2000): 775–795.

12. Branas et al., "A Difference-in-Differences Analysis"; F. E. Kuo and W. C. Sullivan, "Environment and Crime in the Inner City: Does Vegetation Reduce Crime?," *Environment and Behavior* 33 (2001): 343–367; F. E. Kuo et al., "Fertile Ground for Community: Inner-City Neighborhood Common Spaces," *American Journal of Community Psychology* 26, no. 6 (1998): 823–855.

13. Studies of Robert Taylor Homes are also included in chapter 3. See N. Lemann, *The Promised Land: The Great Black Migration and How It Changed America* (New York: Random House, 1991); C. Lewis and S. Sturgill, "Comment: Healing in the Urban Environment: A Person/Plant Viewpoint," *Journal of the American Planning Association* 45, no. 3 (1979): 330–338.

14. According to Kuo et al. (1998), "compared to residents living adjacent to barren spaces, individuals living adjacent to greener common spaces had more social activities and more visitors, knew more of their neighbors, reported their neighbors were more concerned with helping and supporting one another, and had stronger feelings of belonging." From Kuo et al., "Fertile Ground for Community," 843. Being around trees in low-income urban neighborhoods has a number of additional social benefits for residents including greater sense of safety and lower levels of fear and aggressive and violent behavior. See Kuo and Sullivan, "Environment and Crime in the Inner City"; F. E. Kuo, M. Bacaicoa, and W. C. Sullivan, "Transforming Inner-City Landscapes: Trees, Sense of Safety, and Prefer-

ence," *Environment and Behavior* 30 (1998): 28–59; W. Sullivan, F. Kuo, and S. F. DePooter, "The Fruit of Urban Nature: Vital Neighborhood Spaces," *Environment and Behavior* 36, no. 5 (2004): 678–700.

15. Additional cognitive benefits of spending time in nature encompass improved concentration and working memory, intellectual functioning, and attention, including reduced symptoms of Attention Deficit Hyperactive Disorder (ADHD). For reviews of studies on cognitive and other benefits of nature within a "red zone" greening context, see H. Okvat and A. Zautra, "Sowing Seeds of Resilience: Community Gardening in a Post-Disaster Context," in *Greening in the Red Zone*, ed. K. G. Tidball and M. E. Krasny (New York: Springer, 2014), 73–90; Wells, "The Role of Nature in Children's Resilience."

16. E. K. Nisbet and J. M. Zelenski, "Underestimating Nearby Nature: Affective Forecasting Errors Obscure the Happy Path to Sustainability," *Psychological Science* 22, no. 9 (2011): 1101–1106. For a study of how being connected to nature is associated with happiness, see J. M. Zelenski and E. K. Nisbet, "Happiness and Feeling Connected: The Distinct Role of Nature Relatedness," *Environment and Behavior* (2012): 3–23.

17. Relf, "Therapeutic Values of Plants."

18. R. Kaplan and S. Kaplan, "Preference, Restoration, and Meaningful Action in the Context of Nearby Nature," in *Urban Place: Reconnecting with the Natural World*, ed. P. F. Barlett, 271–298 (Cambridge, MA: MIT Press, 2005). The Kaplans call their theory about how nature helps to restore our ability to concentrate Attention Restoration Theory. They also have proposed the Reasonable Person Model, which suggests that people are more healthy psychologically, and thus more reasonable, if they have opportunities to explore new environments and learn new information, act in meaningful ways (e.g., volunteer to help others), and experience the restorative value of nature. S. Kaplan and R. Kaplan, "Health, Supportive Environments, and the Reasonable Person Model," *American Journal of Public Health* 93, no. 9 (2001): 1484–1489.

19. B. J. Park et al., "The Physiological Effects of Shinrin-Yoku (Taking in the Forest Atmosphere or Forest Bathing): Evidence from Field Experiments in 24 Forests across Japan," *Environmental Health and Preventive Medicine* 15 (2010): 18–26.

20. Ulrich, "View through a Window"; Okvat and Zautra, "Sowing Seeds of Resilience"; P. H. Kahn Jr. et al., "A Plasma Display Window?—the Shifting Baseline Problem in a Technologically Mediated Natural World," *Journal of Environmental Psychology* 28 (2008): 192–199.

21. Other writers have talked about similar outcomes of well-being from engaging in environmental stewardship and restoration. Related to the notion of generativity is a sense of satisfaction from engaging in meaningful action. See I. Miles, W. Sullivan, and F. Kuo, "Ecological Restoration Volunteers: The Benefits of Participation," *Urban Ecosystems* 2 (1998): 27–41. Related to generativity and self-efficacy are a sense of pride and of competence that leads to further participation in neighborhood improvement: M. E. Austin and R. Kaplan, "Identity, Involvement, and Expertise in the Inner City: Some Benefits of Tree-Planting Proj-

ects," in *Identity and the Natural Environment: The Psychological Significance of Nature*, ed. S. Clayton and S. Opotow (Cambridge, MA: MIT Press, 2003), 205–225; Kaplan and Kaplan, "Preference, Restoration, and Meaningful Action."

22. Erik Erikson and colleagues talk about the notion of grand-generativity in the latter stages of life as follows. "The capacity for grand-generativity incorporates care for the present with concern for the future—for today's younger generations in their futures, for generations not yet born, and for the survival of the world as a whole. It contributes to the sense of immortality that becomes so important in the individual's struggle to transcend realistic despair as the end of life approaches, inevitably." E. H. Erikson, J. M. Erikson, and H. Q. Kivnick, *Vital Involvement in Old Age* (New York: W. W. Norton & Company, 1986), 74–75. Whereas Erikson focuses on the relationships of aging parents to their children, Warburton and Gooch make the connections between Erikson's generativity and leaving an environmental legacy through volunteer environmental stewardship. Warburton, J., and M. Gooch. "Stewardship Volunteering by Older Australians: The Generative Response." *Local Environment* 12, no. 1 (2007): 43–55.

23. Warburton and Gooch, "Stewardship Volunteering by Older Australians," 43–55.

24. Ibid., 48.

25. Bandura, A. "Perceived Self-Efficacy in Cognitive Development and Functioning," *Educational Psychologist* 28, no. 2 (1993): 117–148.

26. Albert Bandura claims that self-efficacy is formed through four processes: performance accomplishment, vicarious experience, verbal persuasion, and physiological states. Civic ecology would focus primarily on performance accomplishments, which Bandura links to personal mastery experiences. "Self-Efficacy: Toward a Unifying Theory of Behavioral Change," *Psychological Review* 84, no. 2 (1977): 191–215.

27. Bandura, "Perceived Self-Efficacy," 118.

28. Ibid., 133.

29. C. C. Benight and A. Bandura, "Social Cognitive Theory of Posttraumatic Recovery: The Role of Perceived Self-Efficacy," *Behaviour Research and Therapy* 42 (2004): 1129–1148.

30. According to Benight and Bandura, self-efficacy in harsh environments (coping efficacy) is "accompanied by benign appraisals of potential threats, weaker stress reactions to them, less ruminative preoccupation with them, better behavioral management of threats, and faster recovery of well-being from any experienced distress over them. These enabling and protective efficacy-governed processes help to weaken the enduring distressfulness and debilitativeness of traumatizing events" (ibid., 1133). They go on to talk about how human agency plays a role in trauma situations.

31. Bandura, "Self-Efficacy."

32. In addition to modeling and guiding others in mastering particular skills, more experienced individuals provide incentives for others to engage in benefi-

cial activities, and motivate others by demonstrating that perseverance overcomes obstacles. Ibid.

33. L. Saldivar and M. E. Krasny. "The Role of NYC Latino Community Gardens in Community Development, Open Space, and Civic Agriculture." *Agriculture and Human Values* 21 (2004): 399–412. See also the steward stories of Nilka Martell and Andre Rivera in the Bronx.

34. J. Atkinson, "An Evaluation of the Gardening Leave Project for Ex-Military Personnel with PTSD and Other Combat Related Mental Health Problems" (Glasgow, UK: The Pears Foundation, 2009). See also a short video demonstrating how participating in salmon restoration empowered and engendered hope in U.S. veterans, NOAA, "Military Veterans Help Rebuild Fisheries," May 22, 2013, http://www.habitat.noaa.gov/highlights/vetsrestorationvideo.html.

35. J. Rappaport, "In Praise of Pardox: A Social Policy of Empowerment over Prevention," *American Journal of Community Psychology* 9, no. 1 (1981): 1–25.

36. An empowerment ideology "demands that we look to many diverse local settings where people are already handling their own problems in living, in order to learn more about how they do it." Ibid., 15.

37. Although similar to self-efficacy in that it implies control over one's life, empowerment incorporates the notion of power, including granting and sharing of power, in ways that self-efficacy does not. Empowerment is "a multi-dimensional social process that helps people gain control over their own lives. It is a process that fosters power (that is, the capacity to implement) in people, for use in their own lives, their communities, and in their society, by acting on issues that they define as important." N. Page and C. E. Czuba, "Empowerment: What Is It?," *Journal of Extension* 37, no. 5 (1999), http://www.joe.org/joe/1999october/comm1.php. Inherent to empowerment is sharing power, although this does not always happen; for example, community greening organizers in Chicago assumed but did not share control and decision making with other participants, leading to limited engagement of participants who were not in control. L. M. Westphal, "Urban Greening and Social Benefits: A Study of Empowerment Outcomes," *Journal of Arboriculture* 29, no. 3 (2003): 137–147.

38. S. M. Stuart, "Lifting Spirits: Creating Gardens in Domestic Violence Shelters," in *Urban Place: Reconnecting with the Natural World*, ed. P. F. Barlett (Cambridge, MA: MIT Press, 2005), 61–88.

39. J. Schilling and J. Logan, "Greening the Rust Belt: A Green Infrastructure Model for Right Sizing America's Shrinking Cities," *Journal of the American Planning Association* 74, no. 4 (2008): 451–466.

40. R. J. Slater, "Urban Agriculture, Gender and Empowerment: An Alternative View." *Development Southern Africa* 18, no. 5 (2001): 635–650.

41. A *spaza* shop is an informal convenience shop business in South Africa, usually run from home. http://en.wikipedia.org/wiki/Spaza_shop.

42. *Hokkie* means shack.

43. Slater, "Urban Agriculture," 648.

44. In this and other instances, community greening addresses issues related to empowerment of women, immigrants, internal migrants, and other traditionally less powerful members of society. For an account of female gardeners and empowerment in post-war Liberia, see E. Moore, "Growing Hope: How Urban Gardens Are Empowering War-Affected Liberians and Harvesting a New Generation of City Farmers," in *Greening in the Red Zone*, ed. Tidball and Krasny, 411–417. For a description of "emancipation" of female immigrants through greening in The Netherlands, see A. E. J. Wals and M. E. van der Waal, "Sustainability-Oriented Social Learning in Multi-Cultural Urban Areas: The Case of the Rotterdam Environmental Centre," in *Greening in the Red Zone*, 379–396.

45. Two caveats are important. First, the authors of the ash tree study did not state that ash tree deaths caused human deaths, but rather that an increase in human deaths was associated with or occurred in the same counties as those where ash trees had died. Further, we are simply proposing one hypothesis for the increase in human mortality from cardiac-related illness. Other explanations are possible.

46. The American chestnut, once a dominant tree throughout eastern U.S. forests, was killed off by a blight in the early twentieth century. A century later, citizen groups are working with scientists to help plant and monitor the growth of chestnut trees bred for resistance to the blight. Citizen task forces in New York State are helping their communities adapt to the loss of ash trees, and one day could play a role in reintroducing ash trees that are resistant to the emerald ash borer. In this way, civic ecology practices might play a role in future recovery of the ash.

47. Cornell Cooperative Extension, "New York Invasive Species Clearinghouse," 2014, http://www.nyis.info/index.php?action=eab.

## Steward Story: Ben Haberthur

1. National Audubon Society, "Toyota Together Green: Ben Haberthur," n.d., http://www.togethergreen.org/fellows/fellow/benjamin-haberthur.

2. A. Aguilar, "Kane County Veteran Helps Younger Vets Find Peace through Nature," *Chicago Tribune*, October 25, 2012.

3. National Audubon Society, "Toyota Together Green."

4. Aguilar, "Kane County Veteran Helps Younger Vets Find Peace through Nature."

## 7   Learning like Bees

1. T. D. Seeley, *Honeybee Democracy* (Princeton: Princeton University Press, 2010).

2. Ibid., 6.

3. A. E. J. Wals, N. van der Hoeven, and H. Blanken, *The Acoustics of Social Learning: Designing Learning Processes That Contribute to a More Sustain-*

*able World* (Wagengingen, The Netherlands: Wagengingen Academic Publishers, 2009).

4. Ibid.

5. Resource dilemmas arise when the following conditions exist: multiple, interdependent stakeholders make competing claims about a common pool resource; complexity, due to multiple causes and effects and challenges in defining and monitoring "the problem"; and uncertainty. C. Blackmore, "What Kinds of Knowledge, Knowing and Learning Are Required for Addressing Resource Dilemmas?: A Theoretical Overview," *Environmental Science and Policy* 10 (2007): 513.

6. For discussion of importance of multiple forms of knowledge in resource management, see F. Berkes and N. J. Turner, "Knowledge, Learning and the Evolution of Conservation Practice for Social-Ecological System Resilience," *Human Ecology* 34, no. 4 (2006): 479–494; I. Davidson-Hunt and M. O'Flaherty, "Researchers, Indigenous Peoples, and Place-Based Learning Communities," *Society & Natural Resources* 20 (2007): 291–305.

7. Bandura defined social learning as learning through imitation (A. Bandura, *Social Learning Theory* (Englewood Hills, NJ: Prentice-Hall, Inc., 1977). Since that time, his initial considerations of how people learn through interacting with each other has been greatly expanded. See M. Muro and P. Jeffrey, "A Critical Review of the Theory and Application of Social Learning in Participatory Natural Resource Management Processes," *Journal of Environmental Planning and Management* 51, no. 3 (2008): 325–344.

8. J. Lave and E. Wenger, *Situated Learning* (Cambridge, UK: Cambridge University Press, 1991); E. Wenger, R. McDermott, and W. M. Snyder, *Cultivating Communities of Practice* (Cambridge, MA: Harvard Business School Press, 2002).

9. W. K. Stevens, *Miracle under the Oaks: The Revival of Nature in America* (New York: Pocket Books, 1995); D. K. Moskovits et al., "Chicago Wilderness: A New Force in Urban Conservation," *Annals of the New York Academy of Sciences* 1023 (2004): 215–236.

10. M. E. Krasny and K. G. Tidball, "Community Gardens as Contexts for Science, Stewardship, and Civic Action Learning," *Cities and the Environment* 2, no. 1 (2009). http://digitalcommons.lmu.edu/cgi/viewcontent. cgi?article=1037&context=cate; A. M. Kennedy and M. E. Krasny, "Garden Mosaics: Connecting Science to Community," *The Science Teacher* 72, no. 3 (2005): 44–48.

11. In this way, the youth are helping build social capital. For a discussion of social capital in youth environmental education and stewardship programs, see chapter 3 and M. E. Krasny et al., "Measuring Social Capital among Youth: Applications in Environmental Education," *Environmental Education Research*, 2013, http://www.tandfonline.com/doi/pdf/10.1080/13504622.2013.84364.

12. Rocking the Boat (RTB), n.d., http://www.rockingtheboat.org.

13. The notions of affordances and abilities put forth by James Gibson may be useful here. J. J. Gibson, "The Theory of Affordances," in *Perceiving, Acting, and*

*Knowing: Toward an Ecological Psychology,* ed. R. Shaw and J. Bransford (Hillsdale: Lawrence Erlbaum Associates, 1977), 67–82. James Greeno has described affordances as follows: "In any interaction involving an agent with some other system, conditions that enable that interaction include some properties of the agent along with some properties of the other system. . . . The term *affordance* refers to whatever it is about the environment that contributes to the kind of interaction that occurs. One also needs a term that refers to whatever it is about the agent that contributes to the kind of interaction that occurs. I prefer the term *ability* . . . (although others use the terms effectivity or aptitude). . . . Neither an affordance nor an ability is specifiable in the absence of specifying the other. It does not go far enough to say that an ability depends on the context of environmental characteristics, or that an affordance depends on the context of an agent's characteristics. The concepts are codefining, and neither of them is coherent, absent the other." From J. G. Greeno, "Gibson's Affordances," *Psychological Review* 101, no. 2 (1994): 338. See also L. Chawla, "Learning to Love the Natural World Enough to Protect It."

14. J. G. Greeno, "Gibson's Affordances," *Psychological Review* 101, no. 2 (1994): 336–342; Chawla, "Learning to Love the Natural World Enough to Protect It."

15. S. A. Barab and W.-M. Roth, "Curriculum-Based Ecosystems: Supporting Knowing from an Ecological Perspective," *Educational Researcher* 35, no. 5 (2006): 3–13; Chawla, "Participation and the Ecology of Environmental Awareness" in *Participation and Learning: Perspectives on Education and the Environment, Health and Sustainability,* ed. A. Reid, B. B. Jensen, and J. Nikel, 98–110. (New York: Springer-Verlag, 2008); J. G. Greeno, "Situativity of Knowing, Learning, and Research," *American Psychologist* 53, no. 1 (1998): 5–26; K. G. Tidball and M. E. Krasny, "Toward an Ecology of Environmental Education and Learning," *Ecosphere* 2, no. 2 (2011), http://www.esajournals.org/doi/pdf/10.1890/ES10-00153.1.

16. Ideas about learning as an ongoing process, through which individuals as well as the larger social-ecological systems of which they are a part change, are supported by the work of psychologists like Stanford University's James Greeno and Yrjö Engeström at the University of Helsinki. Greeno describes how learning is situated within the system of which the learner is a part, which he refers to as a "situative perspective" on learning, stating, "the main distinguishing characteristic of the situative perspective is its theoretical focus on interactive systems that are larger than the behavior and cognitive processes of an individual agent." Situations that lend themselves to analysis "at the level of interactive systems rather than individual agents" include emergent problem spaces, interactive construction of understanding, and engagement in activities that contribute to group functions and individual identity. See Greeno, "Situativity of Knowing," 5–7.

Similar to Gibson's affordances and situativity, Engeström's activity theory suggests that learning occurs through the interaction of the learner with other components of the system surrounding her, in this case an activity system that includes the tools the learner has at her disposal and the people with whom she interacts. Y. Engeström, ed., *Learning by Expanding: An Activity-Theoretical Approach to Developmental Research* (Helsinki: Orienta-Konsultit, 1987). Interestingly, in language that echoes notions used in discussing resource dilemmas and complex social-ecological systems (see chapter 9, this book), actions within activity systems are "characterized by ambiguity, surprise, interpretation, sense making, and potential for change." Y. Engeström, "Expansive Learning at Work: Toward an Activity-Theoretical Conceptualization," *Journal of Education and Work* 14, no. 1 (2001): 134. For an application of activity theory in a civic ecology learning context, see M. E. Krasny and W.-M. Roth, "Environmental Education for Social-Ecological System Resilience: A Perspective from Activity Theory," *Environmental Education Research* 16, no. 5–6 (2010): 545–558; Tidball and Krasny, "Toward an Ecology of Environmental Education and Learning."

17. Reef Ball Foundation, "The Reef Ball Foundation-Designed Artificial Reefs," http://www.reefball.org.

18. D. Monti, "Experimental Reefs Could Add Life to Bay: No Fluke," December 11, 2013, http://www.warwickonline.com/stories/Experimental-reefs-could-add-life-to-Bay,88203?print=1.

19. Eternal Reefs Inc., "Eternal Reefs," 2014, http://www.eternalreefs.com/.

20. D. A. Fahrenthold, "Last House on Sinking Chesapeake Bay Island Collapses," *Washington Post*, October 26, 2010. http://www.washingtonpost.com/wp-dyn/content/article/2010/10/24/AR2010102402996.html.

21. L. E. Harris, "Artificial Reefs for Ecosystem Restoration and Coastal Erosion Protection with Aquaculture and Recreational Amenities," 2006, http://www.reefball.org/album/%3D%3D%29%20Non-Geographic%20defined%20Photos/artificialreefscientificpapers/2006JulyLEHRBpaper.pdf.

22. C. Pahl-Wostl, "The Importance of Social Learning in Restoring the Multifunctionality of Rivers and Floodplains," *Ecology and Society* 1, no. 1 (2006), http://www.ecologyandsociety.org/vol11/iss1/art10/.

23. S. Gough and W. Scott, "Promoting Environmental Citizenship through Learning: Toward a Theory of Change," in *Environmental Citizenship: Getting from Here to There*, ed. A. Dobson and D. Bell (Cambridge MA: MIT Press, 2006), 275.

24. Stevens, *Miracle under the Oaks.*

25. Speech interrupters in quotes removed. J. E. Delia, "Cultivating a Culture of Authentic Care in Urban Environmental Education: Narratives from Youth Interns at East New York Farms!," M.S. thesis (Ithaca, NY: Cornell University, 2013), 70.

26. Ibid., 47.

27. A. Kudryavtsev, "Sense of Place in Urban Environmental Education," Ph.D. diss. (Ithaca, NY: Cornell University, 2013), 212, http://dx.doi.org/1813/34149.

28. Ibid., 213.

## Steward Story: Andre Rivera

1. Story compiled from a phone interview with Andre Rivera, February 24, 2014. See also A. Kudryavtsev, "Urban Environmental Education and Sense of Place," Ph.D. diss. (Ithaca, NY: Cornell University, 2012), http://dx.doi.org/1813/34149; L. Merino, "Andre's River," 2012, http://vimeo.com/49071323.

2. Kudryavtsev, "Urban Environmental Education and Sense of Place," 201.

3. NYC Parks, "Concrete Plant Park," n.d., http://www.nycgovparks.org/park -features/concrete-plant-park/planyc.

## 8    Governance

1. As quoted in A. W. Hopkins, *Groundswell: Stories of Saving Places, Finding Community* (San Francisco: The Trust for Public Land, 2005).

2. Description compiled from memories of a visit to the Alabama Community Garden by Marianne Krasny and from information contributed by Houston Urban Harvest Board of Directors member Betty Baer. See also K. Huber, "Alabama Grows More Than Veggies: It Grows Spirit," *Houston Chronicle*, March 10, 2007, http://www.chron.com/life/gardening/article/Alabama-grows-more-than-veggies -It-grows-spirit-1832754.php.

3. Gerry Stoker outlines five principles defining governance:

1. Governance refers to a set of institutions and actors that are drawn from but also beyond government.

2. Governance identifies the blurring of boundaries and responsibilities for tackling social and economic issues.

3. Governance identifies the power dependence involved in the relationships between institutions involved in collective action.

4. Governance is about autonomous self-governing networks of actors.

5. Governance recognizes the capacity to get things done which does not rest on the power of government to command or use its authority. It sees government as able to use new tools and techniques to steer and guide. (G. Stoker, "Governance as Theory: Five Propositions," *International Social Sciences Journal* 50, no. 155 [1998]: 18)

Others have extended Stoker's second principle to encompass environmental problems. Environmental governance focuses on complex problems that cross political boundaries, and whose solutions require government, NGO, business, and civil society cooperation at local regional, national, and international levels. United Nations Environment Programme, "Environmental Governance," 2009, http://www.unep.org/pdf/brochures/EnvironmentalGovernance.pdf.

In language that mirrors notions about feedback, change, and adaptation used in this book (see chapter 9), Stoker also states: "One of the difficulties of identifying an organizing perspective that is devoted to understanding a changing system of governance is that no sooner is the perspective outlined than the object of study changes. It is to be hoped, therefore, that the governance perspective can develop in an evolutionary way to capture the processes of adaptation, learning and experiment that are characteristic of governance." Stoker, "Governance as Theory," 26.

4. City of Houston, "Houston Community Gardening Program," 2014. http:// www.houstontx.gov/health/Community/garden.html; Houston Parks and Recreation Department, "Current Community Gardens," 2014, http://www.houstontx. gov/parks/urbangardener/urbangardener_Current.html; Urban Harvest, "Putting Down Roots across the City," 2014, http://urbanharvest.org/communitygardens; U.S. Department of Agriculture, "The People's Garden," 2014, http://www.usda. gov/wps/portal/usda/usdahome?navid=PEOPLES_GARDEN; City of Houston, "Green Houston," 2014, http://www.greenhoustontx.gov; CEC, "Citizens' Environmental Coalition," 2014, http://www.cechouston.org; ICLEI, "URBIS—Urban Biosphere Initiative," 2014, http://urbis.iclei.org.

Similar to the multiple collaborations in the Houston community garden example, seventy-six different groups partnered in the Bronx River Project in New York. Hopkins, *Groundswell*. See also Groundwork-Anacostia and the Urban Waters Federal Partnership in "The Principled Chapter."

5. E. Ostrom, "Polycentric Systems for Coping with Collective Action and Global Environmental Change," *Global Environmental Change* 20 (2010): 550–557.

6. See A. Kudryavtsev, R. C. Stedman, and M. E. Krasny, "Sense of Place in Environmental Education," *Environmental Education Research* 18, no. 2 (2011): 229–250; Y.-F. Tuan, *Topophilia* (Englewood Cliffs, NJ: Prentice-Hall, 1974).

7. B. H. Walker et al., "A Handful of Heuristics and Some Propositions for Understanding Resilience in Social-Ecological Systems," *Ecology and Society* 11, no. 1 (2006), http://www.ecologyandsociety.org/vol11/iss1/art13/.

8. Civic Ecology Lab researcher Bryce Dubois was engaged in these activities and contributed this information.

9. E. Svendsen and L. Campbell, "Land-Markings: 12 Journeys through 9/11 Living Memorials" (Newtown Square, PA: U.S. Department of Agriculture, Forest Service, Northern Research Station, 2006), http://www.livingmemorialsproject .net/DOWNLOADS/NRS-INF-1-06.pdf.

10. K. G. Tidball et al., "Stewardship, Learning, and Memory in Disaster Resilience," *Environmental Education Research* 16, nos. 5–6 (2010): 591–609.

11. Ostrom, "Polycentric Systems," 552.

12. Relative to governance of common property resources: "Institutional arrangements must be complex, redundant, and nested in many layers. . . . Simple strategies for governing the world's resources that rely exclusively on imposed markets or one-level, centralized command and control and that eliminate apparent redundancies in the name of efficiency have been tried and have failed. Catastrophic failures often have resulted when central governments have exerted sole authority over resources." T. Dietz, E. Ostrom, and P. C. Stern, "The Struggle to Govern the Commons," *Science* 302, no. 5652 (December 12, 2003): 1910.

13. E. P. Weber, *Bringing Society Back In: Grassroots Ecosystem Management, Accountability, and Sustainable Communities* (Cambridge, MA: MIT Press, 2003).

14. Our description of the Los Angeles River case draws heavily from quotes, media reports, and the analysis in: R. Gottlieb and A. Azuma, "Re-envisioning the Los Angeles River: An NGO and Academic Institute Influence the Policy Discourse," *Golden Gate University Law Review* 35, no. 3 (2005): 321–342.

15. Gottlieb and Azuma, "Re-envisioning the Los Angeles River," 4.

16. Ibid.

17. Ibid., 14. See also R. Gottlieb, *Reinventing Los Angeles* (Cambridge, MA: MIT Press, 2007).

18. Gottlieb and Azuma, "Re-envisioning the Los Angeles River," 334.

19. FoLAR, "Friends of the Los Angeles River," n.d., http://folar.org; Gottlieb and Azuma, "Re-envisioning the Los Angeles River."

20. D. John, "Civic Environmentalism," in *Environmental Governance Reconsidered*, ed. R. F. Durant, D. Fiorini, and R. O'Leary, 219–254 (Cambridge, MA: MIT Press, 2004); C. Sirianni, *Investing in Democracy: Engaging Citizens in Collaborative Governance* (Washington, DC: Brookings Institution Press, 2009).

21. A social entrepreneur is "someone who recognizes a social problem and uses entrepreneurial principles to organize, create, and manage an initiative to bring about social change." R. Biggs, F. Westley, and S. Carpenter, "Navigating the Back Loop: Fostering Social Innovation and Transformation in Ecosystem Management," *Ecology and Society* 15, no. 2 (2010), http://www.ecologyandsociety.org /vol15/iss2/art9/.

22. G. Mulgan, "Social Innovation: What It Is, Why It Matters and How It Can Be Accelerated" (London: Young Foundation, 2006), 6. For more on social and policy innovations, see chapter 10.

23. C. Reed, "The Ecological (and Urbanistic) Agency of Infrastructure," in *Deconstruction/Reconstruction: The Cheonggyecheon Restoration Project in Seoul*, ed. J. Busquets (Cambridge, MA: Harvard University Graduate School of Design, 2011), 35–46

24. E. Brenner, "Restored River a Boon to Yonkers," *New York Times*, August 9, 2012. http://www.nytimes.com/2012/08/12/realestate/westchester-in-the-region -restored-river-a-boon-to-yonkers.html.

25. F. A. Bernstein, "A Vision of a Park on a Restored Los Angeles River," *New York Times*, September 28, 2010. http://www.nytimes.com/2010/09/29/ realestate/29river.html?_r=1.

26. D. John, "Civic Environmentalism"; C. Sirianni, *Investing in Democracy*.

27. Biggs, Westley, and Carpenter "Navigating the Back Loop."

28. Dietz, Ostrom, and Stern, "Struggle to Govern the Commons," 1908.

## Steward Story: Nam-Sun Park

1. Story compiled from an interview conducted and translated by Eunju Lee, at Donghae city in Kangwon province, January 19, 2014.

2. ICLEI, "FAQ: ICLEI, the United Nations, and Agenda 21," n.d., http://www.icleiusa.org/about-iclei/faqs/faq-iclei-the-united-nations-and-agenda-21-what-is-agenda-21.

## 9 Resistance, Remembrance, Revolt—and Resilience

1. Anon., "Conservationist Found Murdered in Cape Town Home," *Grassroots: Newsletter of the Grassland Society of Southern Africa* 8, no. 1 (February 2008): 2, http://grassland.org.za/resources/publications/grassroots/2008/February%20 2008/1%20Regulars%20Feb%202008.pdf/view.

2. D. Meadows and P. Marshall, "Dancing with Systems," *Whole Earth* 106 (2001): 63.

3. M. Bergman, "Athlone History: The Trojan Horse Massacre," n.d., http://www.athlone.co.za/heritage/history.php.

4. For a description of virtuous cycles in civic ecology contexts, see K. G. Tidball and M. E. Krasny, "Toward an Ecology of Environmental Education and Learning," *Ecosphere* 2, no. 2 (2011), http://www.esajournals.org/doi/pdf/10.1890/ES10-00153.1.

5. B. Pitt and T. Boulle, "Growing Together: Thinking and Practice of Urban Nature Conservators" (Cape Town: SANBI Cape Flats Nature, 2010).

6. Ibid., chapter 4, 22.

7. O. O'Brien, "Cape Flats, Cape Town," May 24, 2012, http://blogs.newschool.edu/epsm/2012/05/24/cape-flats-cape-town/.

8. F. Berkes, J. Colding, and C. Folke, *Navigating Social-Ecological Systems: Building Resilience for Complexity and Change* (Cambridge, UK: Cambridge University Press, 2003). The U.S. National Science Foundation and other well-known scholars use the term "coupled human and natural systems" to emphasize the importance of integrating human and ecosystem perspectives in addressing environmental problems. See J. Liu et al., "Complexity of Coupled Human and Natural Systems," *Science* 317 (2007): 1513–1516. We prefer the term "social-ecological systems" because the language is less likely to suggest that humans are separate from "nature." Scholars have used the concept of social-ecological systems to emphasize that humans are integrated as part of nature and to stress that the delineation between social systems and ecological systems is artificial and arbitrary. Further, the social-ecological systems approach holds that social and ecological systems are linked through feedback mechanisms and that it is imperative to study them both together in order to understand resilience and complexity. Berkes, Colding, and Folke, *Navigating Social-Ecological Systems.*

9. L. H. Gunderson and C. S. Holling, *Panarchy: Understanding Transformations in Human and Natural Systems* (Washington, DC: Island Press, 2002); C. S. Holling, "The Spruce-Budworm/Forest-Management Problem," in *Adaptive Environmental Assessment and Management*, ed. C. S. Holling (Hoboken: John Wiley & Sons, 1978), 143–182.

10. A powerful example of such a window of opportunity was the decline of the apartheid system in South Africa and Nelson Mandela emerging from prison to take the reins of a pluralistic democracy. Additional opportunities appeared in the conservation and civic sectors in South Africa. For example, the window opened for black and colored South Africans to partner with government conservation agencies, resulting in novel resource management approaches referred to as "pockets of social-ecological innovation." H. Ernstson and T. Elmqvist, "Globalization, Urban Ecosystems and Social-Ecological Innovations: A Comparative Network Analytic Approach" (Stockholm: Stockholm University, 2011).

11. Gunderson and Holling, *Panarchy*.

12. F. Berkes and C. Folke, "Back to the Future: Ecosystem Dynamics and Local Knowledge, " in *Panarchy: Understanding Transformations in Human and Natural Systems*, ed. L. H. Gunderson and C. S. Holling, 121–146 (Washington, DC: Island Press, 2002); C. Folke et al., "Resilience and Sustainable Development: Building Adaptive Capacity in a World of Transformations" (Johannesburg, South Africa: World Summit on Sustainable Development, 2002), http://www .resalliance.org/files/1144440669_resilience_and_sustainable_development.pdf; J. A. Tainter, "Global Change, History, and Sustainability," in *The Way the Wind Blows: Climate, History, and Human Action*, ed. R. J. McIntosh, J. A. Tainter, and S. K. McIntosh, 331–356 (New York: Columbia University Press, 2000).

13. C. Folke et al., "Resilience Thinking: Integrating Resilience, Adaptability and Transformability," *Ecology and Society* 15, no. 4 (2010), http://www.ecologyand society.org/vol15/iss4/art20/.

14. S. Bowles, S. N. Durlauf, and K. Hoff, *Poverty Traps* (Princeton: Princeton University Press, 2006); J. D. Sachs, *The End of Poverty: Economic Possibilities for Our Time* (New York: Penguin Press, 2005). For a discussion of poverty and rigidity traps and adaptive capacity, see S. R. Carpenter and W. A. Brock, "Adaptive Capacity and Traps," *Ecology and Society* 13, no. 2 (2008), http://www.ecolo gyandsociety.org/vol13/iss2/art40/.

15. M.-L. Moore and F. Westley, "Surmountable Chasms: Networks and Social Innovation for Resilient Systems," *Ecology and Society* 16, no. 1 (2011), http:// www.ecologyandsociety.org/vol16/iss1/art5/.

16. Ibid.; M. Scheffer and F. Westley, "The Evolutionary Basis of Rigidity: Locks in Cells, Minds, and Society," *Ecology and Society* 12, no. 2 (2007), http://www .ecologyandsociety.org/vol12/iss2/art36/.

17. Ashden Awards, "Trees, Water and People (TWP) and Asociación Hondureña Para El Desarrollo (AHDESA), Honduras," December 2009, http://www.ashden.org /files/TWP full.pdf; E. Díaz et al., "Self-Rated Health among Mayan Women Participating in a Randomised Intervention Trial Reducing Indoor Air Pollution in Guatemala," *BMC International Health and Human Rights* 8, no. 1 (2008): 1–8.

18. L. P. Jackson, "Storm Fuels Environmental Awareness: A Guest Column by Lisa P. Jackson," *The Times-Picayune*, August 26, 2010, http://www.nola.com/katrina/index.ssf/2010/08/storm_fuels_environmental_awar.html. In part due to efforts of New Orleans community members, the BP oil and gas company settlement included not only compensation for immediate loss of livelihood, but also $2.4 billion for coastal restoration. D. Malakoff, "Research, Ecological Restoration Get Unprecedented Windfall from BP Oil Spill Criminal Settlement," *Science News* 2012.

19. K. G. Tidball, "Greening in the Red Zone: Valuing Community-Based Ecological Restoration in Human Vulnerability Contexts," Ph.D. diss (Ithaca, NY: Cornell University, 2012), http://dspace.library.cornell.edu/handle/1813/31450; K. G. Tidball and M. E. Krasny, *Greening in the Red Zone* (New York: Springer, 2014).

20. M. Cramer, "The Berlin Wall Trail: A Cycling and Hiking Route on the Traces of Berlin's East-West Division during the Cold War," in *Greening in the Red Zone*, ed. Tidball and Krasny, 445–449.

21. S. Cheng and J. R. McBride, "Restoration of the Urban Forests of Tokyo and Hiroshima Following World War II," in *Greening in the Red Zone*, ed. Tidball and Krasny, 225–248.

22. D. Winterbottom, "Developing a Safe, Nurturing and Therapeutic Environment for the Families of the Garbage Pickers in Guatemala and for Disabled Children in Bosnia and Herzegovina," in *Greening in the Red Zone*, ed. Tidball and Krasny, 397–410.

23. E. Lee, "Reconstructing Village Groves after a Typhoon in Korea," in *Greening in the Red Zone*, ed. Tidball and Krasny, 159–162.

24. Folke et al., "Resilience and Sustainable Development"; A. Löf, "Exploring Adaptability through Learning Layers and Learning Loops," *Environmental Education Research* (special issue on Resilience in Social-Ecological Systems: The Roles of Learning and Education; ed. M. E. Krasny, C. Lundholm, and R. Plummer) 15, nos. 5–6 (2010): 529–543.

25. Folke et al., "Resilience and Sustainable Development."

26. Ibid., 51.

27. G. P. Richardson, "Systems Dynamics in the Elevator," n.d., http://www.stewardshipmodeling.com/System%2520Dynamics%2520in%2520the%2520Elevator.htm/20Elevator.htm. See also G. P. Richardson, "Problems with Causal Loop Diagrams," *Systems Dynamics Review* 2, no. 2 (1986), http://www.systems-thinking.org/intst/d-3312.pdf.

28. B. H. Walker and D. Salt, *Resilience Thinking: Sustaining Ecosystems and People in a Changing World* (Washington, DC: Island Press, 2006).

29. F. Berkes and C. Folke, *Linking Social and Ecological Systems* (London: Cambridge University Press, 1998); F. Berkes, J. Colding, and C. Folke, "Rediscovery of Traditional Ecological Knowledge as Adaptive Management," *Ecological Applications* 10, no. 5 (2000): 1251–1262.

30. R. Biggs, F. Westley, and S. Carpenter, "Navigating the Back Loop: Fostering Social Innovation and Transformation in Ecosystem Management," *Ecology and Society* 15, no. 2 (2010), http://www.ecologyandsociety.org/vol15/iss2/art9/.

31. S. Levin, "Self-Organization and the Emergence of Complexity in Ecological Systems," *BioScience* 55, no. 12 (2005): 1075–1079.

32. D. Armitage, M. Marschke, and R. Plummer, "Adaptive Co-Management and the Paradox of Learning," *Global Environmental Change* 18 (2008): 86–98; C. Pahl-Wostl et al., "Social Learning and Water Resources Management," *Ecology and Society* 12, no. 2 (2007), http://www.ecologyandsociety.org/vol12/iss2/art5; C. Blackmore, R. Ison, and J. Jiggins, "Social Learning: An Alternative Policy Instrument for Managing in the Context of Europe's Water," *Environmental Science & Policy* 10, no. 6 (2007): 493–498.

33. Meadows and Marshall, "Dancing with Systems."

34. S. S. Luthar, D. Cicchetti, and B. Becker, "The Construct of Resilience: A Critical Evaluation and Guidelines for Future Work," *Child Development* 71, no. 3 (2000): 543–562; G. A. Bonanno, "Loss, Trauma, and Human Resilience: How We Have Underestimated the Human Capacity to Thrive after Extremely Aversive Events," *American Psychologist* 59, no. 1 (2004): 20–28; H. Okvat and A. Zautra, "Sowing Seeds of Resilience: Community Gardening in a Post-Disaster Context," in *Greening in the Red Zone*, ed. Tidball and Krasny, 73–90; A. S. Masten and J. Obradovic, "Disaster Preparation and Recovery: Lessons from Research on Resilience in Human Development," *Ecology and Society* 13, no. 1 (2008), http://www.ecologyandsociety.org/vol13/iss1/art9/.

35. M. E. Krasny et al., "Greening and Resilience in Military Communities," in *Greening in the Red Zone*, ed. Tidball and Krasny, 163–180.

36. E. T. Wimberley, *Nested Ecology: The Place of Humans in the Ecological Hierarchy* (Baltimore: The Johns Hopkins University Press, 2009).

37. C. S. Holling, L. H. Gunderson, and G. D. Peterson, "Sustainability and Panarchies," in *Panarchy*, ed. Gunderson and Holling, 74.

38. Gunderson and Holling, *Panarchy*.

39. Holling, Gunderson, and Peterson, "Sustainability and Panarchies," 76.

40. R. J. McIntosh, J. A. Tainter, and S. K. McIntosh, "Climate, History, and Human Action," in *The Way the Wind Blows: Climate, History, and Human Action*, ed. R. J. McIntosh, J. A. Tainter, and S. K. McIntosh, 1–42 (New York: Columbia University Press, 2000).

41. Holling, Gunderson, and Peterson, "Sustainability and Panarchies," 91.

42. Tidball and Krasny, *Greening in the Red Zone*; Cramer, "The Berlin Wall Trail"; A. Grichting, "Cyprus: Greening in the Dead Zone," in *Greening in the Red Zone*, ed. Tidball and Krasny, 429–443; A. Grichting and K. G. Kim, "The Korea DMZ: From a Red Zone to a Deeper Shade of Green," in *Greening in the Red Zone*, ed. Tidball and Krasny, 197–201.

43. B. Walker et al., "Resilience Management in Social-Ecological Systems: A Working Hypothesis for a Participatory Approach," *Conservation Ecology* 6, no. 1 (2002), http://www.consecol.org/vol6/iss1/art14/.

44. E. D. Weinstein and K. G. Tidball, "Environment Shaping: An Alternative Approach to Development and Aid," *Journal of Intervention and Statebuilding* 1, no. 1 (2007): 67–85; K. G. Tidball, E. D. Weinstein, and M. E. Krasny, "Synthesis and Conclusion: Applying Greening in Red Zones," in *Greening the Red Zone*, ed. Tidball and Krasny, 451–486.

45. Okvat and Zautra, "Sowing Seeds of Resilience."

## Steward Story: Mandla Mentoor

1. M. Lindow, "Mountain of Hope," *Sierra Magazine*, March/April 2005, http://vault.sierraclub.org/sierra/200503/profile.asp.

2. Ibid.

3. Ashoka, "Mandla Mentoor," n.d., https://www.ashoka.org/fellow/mandla-mentoor.

4. S. Shava and M. Mentoor, "Turning Degraded Open Space into a Community Asset: The Soweto Mountain of Hope Greening Case Story," in *Greening in the Red Zone*, ed. K. G. Tidball and M. E. Krasny (New York: Springer, 2014), 91–94.

## 10  Policy Frameworks

1. Friends of Mitchell Park, n.d., http://www.mitchellparkdc.org.

2. D. E. Pataki et al., "Coupling Biogeochemical Cycles in Urban Environments: Ecosystem Services, Green Solutions, and Misconceptions," *Frontiers in Ecology and the Environment* 9, no. 1 (2011): 27–36.

3. Central Park Conservancy, "Official Website of New York City's Central Park," n.d., http://www.centralparknyc.org/about.

4. Ibid.

5. Such multimillion-dollar "outsourcing" can be problematic for those concerned with equity in access to green space. For example, during the recent economic decline, California turned to nonprofit friends of parks associations to manage its state parks. Similar to the Central Park Conservancy, the larger friends associations linked to state parks had a long history and extensive resources, but questions were raised about equity across all parks, including those in low-income neighborhoods. C. Nesper, "Legitimate Protection or Tactful Abandonment: Can Recent California Legislation Sustain the San Francisco Bay Area's Public Lands?," *Golden Gate University Environmental Law Journal* 6, no. 1 (2012): 153–181.

6. J. Ruitenbeek and C. Cartier, "The Invisible Wand: Adaptive Co-management as an Emergent Strategy in Complex Bio-economic Systems," occasional paper (Bogor, Indonesia: Center for International Forestry Research, October 2001),

http://www.cifor.org/publications/pdf_files/OccPapers/OP-034.pdf; E. D. Weinstein and K. G. Tidball, "Environment Shaping: An Alternative Approach to Development and Aid," *Journal of Intervention and Statebuilding* 1, no. 1 (2007): 67–85.

7. Policy entrepreneurs are also referred to as institutional entrepreneurs in the literature cited in this chapter.

8. M.-L. Moore and F. Westley, "Surmountable Chasms: Networks and Social Innovation for Resilient Systems," *Ecology and Society* 16, no. 1 (2011), http://www.ecologyandsociety.org/vol16/iss1/art5/; S. Maguire, C. Hardy, and T. B. Lawrence, "Institutional Entrepreneurship in Emerging Fields: HIV/AIDS Treatment Advocacy in Canada," *The Academy of Management Journal* 47, no. 5 (2004): 657–679; B. Leca, J. Battilana, and E. Boxenbaum, "Agency and Institutions: A Review of Institutional Entrepreneurship," working paper 08-096, Harvard Business School, 2008, http://egateg.usaidallnet.gov/sites/default/files/Review%20of%20Institutional%20Entrepreneurship.pdf; S. Meijerink and D. Huitema, "Policy Entrepreneurs and Change Strategies: Lessons from Sixteen Case Studies of Water Transitions around the Globe," *Ecology and Society* 15, no. 2 (2010), http://www.ecologyandsociety.org/vol15/iss2/art21/; H. Hwang and W. J. Powell, "Institutions and Entrepreneurship," in *The Handbook of Entrepreneurship*, ed. S. A. Alvarez, R. Agrawal, and O. Sorenson (New York: Springer, 2005), 210–232; M. van der Steen and J. Groenewegen, "Exploring Policy Entrepreneurship," in *Coherence between Institutions and Technologies in Infrastructures*, ed. M. Finger and R. Künneke (Rome: European Association for Evolutionary Political Economy, 2008), http://mir.epfl.ch/webdav/site/mir/users/181931/public/wp0801.pdf; T. Heikkila and A. K. Gerlak, "The Formation of Large-Scale Collaborative Resource Management Institutions: Clarifying the Roles of Stakeholders, Science, and Institutions," *Policy Studies Journal* 33, no. 4 (2005): 583–612; N. Fligstein, "Institutional Entrepreneurs and Cultural Frames—the Case of the European Union's Single Market Program," *European Societies* 3, no. 3 (2001): 261–287; Weinstein and Tidball, "Environment Shaping"; R. Biggs, F. Westley, and S. Carpenter, "Navigating the Back Loop: Fostering Social Innovation and Transformation in Ecosystem Management," *Ecology and Society* 15, no. 2 (2010), http://www.ecologyandsociety.org/vol15/iss2/art9/; O. Bodin and B. I. Crona, "The Role of Social Networks in Natural Resource Governance: What Relational Patterns Make a Difference?," *Global Environmental Change* 19 (2009): 366–374.

9. Weinstein and Tidball, "Environment Shaping"; K. G. Tidball, E. D. Weinstein, and M. E. Krasny, "Synthesis and Conclusion: Applying Greening in Red Zones," in *Greening the Red Zone*, ed. K. G. Tidball and M. E. Krasny, 451–486. New York: Springer, 2014.

10. M. E. Krasny et al., Garden Mosaics website, n.d., http://communitygardennews.org/gardenmosaics/.

11. M. E. Krasny et al., "New York City's Oyster Gardeners: Memories, Meanings, and Motivations of Volunteer Environmental Stewards," *Landscape and Urban Planning* 132 (December 2014): 16–25.

12. B. Pitt and T. Boulle, "Growing Together: Thinking and Practice of Urban Nature Conservators" (Cape Town: SANBI Cape Flats Nature, 2010), http://www.capeaction.org.za/index.php/resources/landscape-initiatives?view=category &id=70.

13. K. G. Tidball, "Combat Wounded Veterans Replant Coral Fragments," November 30, 2013, http://reworrr.blogspot.com/2013/11/combat-wounded-veterans-replant-coral.html.

14. E. Lee and M. E. Krasny, "The Role of Social Learning for Social-Ecological Systems in Korean Village Groves Restoration," forthcoming in *Ecology and Society*.

15. J. Longhurst, "Mennonites, Sikhs Work Together on Garden Project," May 26, 2012, http://www.winnipegfreepress.com/arts-and-life/life/faith/mennonites-sikhs-work-together-on-garden-project-planting-seeds-of-harmony-154386235.html.

16. H. Kobori and R. B. Primack, "Participatory Conservation Approaches for *Satoyama*, the Traditional Forest and Agricultural Landscape of Japan," *Ambio* 32, no. 4 (2003): 307–311; R. B. Primack, H. Kobori, and S. Mori, "Dragonfly Pond Restoration Promotes Conservation Awareness in Japan," *Conservation Biology* 14, no. 5 (2000): 1153–1154.

17. N. Tyack, "A Trash Biography" (Los Angeles: Friends of the Los Angeles River, 2011), http://folar.org/wp-contentuploads/2011/11/trashsortreport.pdf.

18. This practice is being implemented by Chester community leaders, including Andrew Brazelton, with support from Akiima Price.

19. Groundwork Hudson Valley, Saw Mill River Coalition, n.d., http://www.sawmillrivercoalition.org.

20. K. G. Tidball, "Trees and Rebirth: Social-Ecological Symbols and Rituals in the Resilience of Post-Katrina New Orleans," in *Greening in the Red Zone*, ed. Tidball and Krasny, 257–296.

21. Meijerink and Huitema, "Policy Entrepreneurs and Change Strategies"; Fligstein, "Institutional Entrepreneurs and Cultural Frames."

22. T. Lyson, *Civic Agriculture: Reconnecting Farm, Food, and Community* (Lebanon, NH: Tufts University Press, 2004).

23. P. R. Ehrlich and H. A. Mooney, "Extinction, Substitution, and Ecosystem Services," *BioScience* 33, no. 4 (1983): 248–254; R. Costanza et al., "The Value of the World's Ecosystem Services and Natural Capital," *Nature* 387 (1997): 253–260.

24. R. Gottlieb and A. Azuma, "Re-envisioning the Los Angeles River: An NGO and Academic Institute Influence the Policy Discourse," *Golden Gate University Law Review* 35, no. 3 (2005): 321–342.

25. P. Olsson et al., "Enhancing the Fit through Adaptive Comanagement: Creating and Maintaining Bridging Functions for Matching Scales in the Kristianstads Vattenrike Biosphere Reserve Sweden," *Ecology and Society* 12, no. 1 (2007). http://www.ecologyandsociety.org/vol12/iss1/art28/.

26. Meijerink and Huitema, "Policy Entrepreneurs and Change Strategies"; Olsson et al., "Enhancing the Fit"; Fligstein, "Institutional Entrepreneurs and Cultural Frames."

27. Olsson et al., "Enhancing the Fit."

28. L. P. Jackson, "Storm Fuels Environmental Awareness: A Guest Column by Lisa P. Jackson." *The Times-Picayune*, August 26, 2010. http://www.nola.com /katrina/index.ssf/2010/08/storm_fuels_environmental_awar.html.

29. K. Mulvaney, "Time to Bring Back New York's Oysters," November 14, 2012, http://news.discovery.com/earth/new-york-needs-new-oysters-121114.htm; A. Revkin, "Students Press the Case for Oysters as New York's Surge Protector," *New York Times*, November 12, 2012, http://dotearth.blogs.nytimes.com /2012/11/12/students-press-the-case-for-oysters-as-new-yorks-surge-protector.

30. M. E. Krasny and K. G. Tidball, "Civic Ecology: A Pathway for Earth Stewardship in Cities," *Frontiers in Ecology and the Environment* 10, no. 5 (2012): 267–273; K. G. Tidball and M. E. Krasny, "From Risk to Resilience: What Role for Community Greening and Civic Ecology in Cities?," in *Social Learning towards a More Sustainable World*, ed. A. E. J. Wals, 149–164. (Wagengingen, The Netherlands: Wagengingen Academic Press, 2007).

31. A. Cornwall and R. Jewkes, "What Is Participatory Research?," *Social Science and Medicine* 41, no. 12 (1995): 1667–1676; W. F. Whyte, ed. *Participatory Action Research* (Thousand Oaks, CA: Sage, 1991); J. Shirk et al., "Public Participation in Scientific Research: A Framework for Deliberate Design," *Ecology and Society* 17, no. 2 (2012), http://www.ecologyandsociety.org/vol17/iss2/art29/.

32. See scientist calls to action including action ecology in K. Marshall et al., "Science for a Social Revolution: Ecologists Entering the Realm of Action," *Bulletin of the Ecological Society of America* 92, no. 3 (2011): 241–243; F. S. Chapin et al., "Earth Stewardship: Science for Action to Sustain the Human-Earth System," *Ecosphere* 2, no. 8 (2011), http://www.esajournals.org/doi/full/10.1890/ ES11-00166.1; M. A. Palmer, "Socioenvironmental Sustainability and Actionable Science," *BioScience* 62, no. 1 (2012): 5–6; W. H. Schlesinger, "Translational Ecology," *Science* 329 (2010): 609; S. Levin and W. Clark, eds., "Toward a Science of Sustainability: Report from Toward a Science of Sustainability Conference," CID Working Paper No. 196 (Cambridge, MA: Center for International Development, Harvard University), May 2010, http://www.hks.harvard.edu/var/ezp_site/stor-age/fckeditor/file/pdfs/centers-programs/centers/cid/publications/faculty/wp/196 .pdf.

33. M. Gregory, "The Garden Ecology Project," January 2011, blogs.cornell.edu /gep/files/2011/02/GEP-Gardener-Update.pdf.

34. A. Kudryavtsev, R. C. Stedman, and M. E. Krasny, "Sense of Place in Environmental Education," *Environmental Education Research* 18, no. 2 (2011): 229–250; A. Kudryavtsev, M. E. Krasny, and R. C. Stedman, "The Impact of Environmental Education on Sense of Place among Urban Youth," *Ecosphere* 3, no. 4 (2012), http://www.esajournals.org/doi/abs/10.1890/ES11-00318.1.

35. L. Briggs and M. E. Krasny, "Conservation in Cities: Linking Citizen Science and Civic Ecology Practices."

36. B. DuBois and K. G. Tidball, "Greening in Coastal New York Post-Sandy," Civic Ecology Lab White Paper Series #CEL004 (Ithaca, NY: Cornell Civic Ecology Lab, Cornell University, 2013), http://civeco.files.wordpress.com/2013/09/2013-dubois.pdf.

37. Another mechanism by which civic ecology practices expand is through the positive feedback between greening and local environmental quality, which can be explored through computer simulations known as agent-based models. See K. G. Tidball, A. Atkipis, and R. C. Stedman, "Greening and Positive Feedback: Cultural, Cognitive and Environmental Factors Influencing Willingness to Invest in Local Ecological Infrastructure through Greening," forthcoming in *Human Ecology*.

38. K. Taylor, "After High Line's Success, Other Cities Look Up," *New York Times*, July 14, 2010, http://www.nytimes.com/2010/07/15/arts/design/15highline.html?_r=1&.

39. P. Olsson, C. Folke, and F. Berkes, "Adaptive Co-Management for Building Resilience in Social-Ecological Systems," *Environmental Management* 34, no. 1 (2004): 75–90; D. Armitage, F. Berkes, and N. Doubleday, "Introduction: Moving Beyond Co-Management," in *Adaptive Co-management: Collaboration, Learning, and Multi-level Governance*, ed. D. Armitage, F. Berkes, and N. Doubleday, (Vancouver: UBC Press, 2007), 1–18; D. Huitema et al., "Adaptive Water Governance: Assessing the Institutional Prescriptions of Adaptive (Co-)Management from a Governance Perspective and Defining a Research Agenda," *Ecology and Society* 14, no. 1 (2009), http://www.ecologyandsociety.org/vol14/iss1/art26/.

40. J. M. Grove, "Cities: Managing Densely Settled Social-Ecological Systems," in *Principles of Ecosystem Stewardship. Resilience-Based Natural Resource Management in a Changing World*, ed. F. S. Chapin, G. P. Kofinas, and C. Folke, 281–294. (New York: Springer, 2009); Moore and Westley, "Surmountable Chasms." "Grassroots innovations" is another term used in the innovation literature, in particular to refer to innovations in practices generated by community groups. G. Seyfang and A. Haxeltine, "Growing Grassroots Innovations: Exploring the Role of Community-Based Initiatives in Governing Sustainable Energy Transitions," *Environment and Planning C: Government and Policy* 30 (2012): 381–400.

41. NYC Department of Parks & Recreation, "The Community Garden Movement: Green Guerrillas Gain Ground," n.d., http://www.nycgovparks.org/about/history/community-gardens/movement.

42. In the community garden and other cases, the role of government is transformed from regulator to "enabler." Government recognizes local assets such as community gardens, as well as the nontangible assets that community gardens bring to the table—including practical knowledge, social networks, and empowered volunteers with a passionate attachment to place. In collaboration with others who have a "stake" in these public goods, governments design policies that support these community-driven efforts. See also C. Sirianni and L. A. Friedland, *Civic Innovation in America: Community Empowerment, Public Policy, and the Movement for Civic Renewal* (Berkeley: University of California Press, 2001); C.

Sirianni and L. A. Friedland, *The Civic Renewal Movement: Community Building and Democracy in the United States* (Dayton, OH: Charles F. Kettering Foundation, 2005).

43. See Moore and Westley, "Surmountable Chasms"; Maguire, Hardy, and Lawrence, "Institutional Entrepreneurship in Emerging Fields"; Leca, Battilana, and Boxenbaum, "Agency and Institutions"; Meijerink and Huitema, "Policy Entrepreneurs and Change Strategies"; Hwang and Powell, "Institutions and Entrepreneurship"; van der Steen and Groenewegen, "Exploring Policy Entrepreneurship"; Heikkila and Gerlak, "Formation of Large-Scale Collaborative Resource Management Institutions"; Fligstein, "Institutional Entrepreneurs and Cultural Frames"; Weinstein and Tidball, "Environment Shaping"; Biggs, Westley, and Carpenter, "Navigating the Back Loop"; Bodin and Crona, "Role of Social Networks."

44. H. Ernstson et al., "Scale-Crossing Brokers and Network Governance of Urban Ecosystem Services: The Case of Stockholm," *Ecology and Society* 15, no. 4 (2010), http://www.ecologyandsociety.org/vol15/iss4/art28/.

45. Moore and Westley, "Surmountable Chasms"; Ernstson et al., "Scale-Crossing Brokers."

46. Meijerink and Huitema, "Policy Entrepreneurs and Change Strategies."

47. R. G. Panetta, *Westchester: The American Suburb* (New York: Fordham University Press, 2006).

48. Y. Alfaro, "The Bronx River Alliance," *The Greener Apple*, December 18, 2009. http://macaulay.cuny.edu/eportfolios/thegreenerapple/2009/12/18/the-bronx-river-alliance/.

49. M. King, "60 Bronx Youths Are on Right Path," *Daily News*, August 18, 1983.

50. L. Campbell, "Civil Society Strategies on Urban Waterways: Stewardship, Contention, and Coalition Building," master's thesis, Massachusetts Institute of Technology, 2006; Alfaro, "Bronx River Alliance"; M. Powell, "West Farms Rapids Park: A History," February 25, 2014, http://bronxriversankofa.wordpress.com/2014/02/25/west-farms-rapids-park-a-history/; A. W. Hopkins, *Groundswell: Stories of Saving Places, Finding Community* (San Francisco: The Trust for Public Land, 2005).

51. Alfaro, "Bronx River Alliance"; M. Goncalves, "Environmentalists Celebrate Return of Alewife," *The Hunts Point Express*, May 28, 2009, http://brie.hunter.cuny.edu/hpe/2009/05/28/little-fish-signal-big-changes-in-bronx-river/.

52. A 2009 article in the local Bronx newspaper *Hunt's Point Express* described how networks of stewardship organizations not only have helped to restore the river, but also have fostered connections among people and the other animals they have brought back. (Alewife is also referred to as river herring.)

Here's what the rebirth of the Bronx River looks like: it's about 10 inches long with silvery skin and it swims.

In late March, naturalists netted four little fish called alewife in the river, and in doing so won a bet they made on the river's health three years ago when 200 adult alewife were freed in the river to spawn. . . .

Once established, the herring will provide food for egrets, herons and hawks, raccoons, river otters and even seals, as well as for larger fish, including striped bass and tuna, . . .

In addition, the return of the alewife will attract fishermen, environmentalists hope. Families that enjoy fishing will become stewards of the river's health . . .

Knowing that the fish mature in three to four years, this spring workers from the Natural Resources Group, Bronx River Alliance and Rocking the Boat staked out strategic points along the river. . . .

They also spotted fish jumping out of the water to snatch bugs from the air. The action is called "popping," said Rachlin. They looked like "silver dollars jumping out of the surface of the water," said Linda Cox, executive director of the Bronx River Alliance.

"The fact that they came back is pretty amazing," said Yuritt Zeron, 19, a student at La Guardia College who started as a volunteer with the alewife program and now works with Rocking the Boat.

Once these fish spawn, they will swim downstream to Long Island Sound and out to sea, never to return again. Their offspring will follow them when they are big enough and will return to the Bronx River to complete the three- to four-year cycle of life. "Now that it is certain the river can support the fish," said Rachlin, "the dams will be breached to create passageways for the next generation of alewife."

The broad coalition engaged in the alewife project includes Lehman College, the Bronx River Alliance, Wildlife Conservation Society, Connecticut Department of Environmental Protection, Rocking the Boat, Sustainable South Bronx, Youth Ministries for Peace and Justice and other community groups, government agencies and nonprofit organizations.

Congressman Serrano, who has poured millions of federal dollars into the effort to clean up the Bronx River, exulted, "The revitalized Bronx River is a symbol for the Bronx itself." The borough, he said, has "come a long way, just like those little fish that we put in our river three years ago." (Goncalves, "Environmentalists Celebrate Return of Alewife")

53. Bronx River Alliance, "History," n.d., http://bronxriver.org/?pg=content&p= aboutus&m1=44. Italics added.

54. Forterra, n.d., http://www.forterra.org.

55. According to its website, the American Community Gardening Association "recognizes that community gardening improves people's quality of life by providing a catalyst for neighborhood and community development, stimulating social interaction, encouraging self-reliance, beautifying neighborhoods, producing nutritious food, reducing family food budgets, conserving resources and creating opportunities for recreation, exercise, therapy and education."

56. EPA, "Urban Waters Federal Partnership," June 13, 2014, http://www.urban waters.gov.

57. ICLEI, "URBIS—Urban Biosphere Initiative," n.d., http://urbis.iclei.org/.

58. Bodin and Crona, "Role of Social Networks."

59. Moore and Westley, "Surmountable Chasms."

60. K. G. Tidball, M. Bain, and T. Elmqvist, "Detecting Virtuous Cycles That Confer Resilience in Disrupted Social-Ecological Systems," forthcoming in *Ecology and Society*.

61. Weinstein and Tidball, "Environment Shaping."

62. G. Mulgan, "Social Innovation: What It Is, Why It Matters and How It Can Be Accelerated" (London: Young Foundation, 2006), http://dsi.britishcouncil.org .cn/images/SocialInnovation.pdf.

## Afterword

1. K. Lovett, "Gov. Cuomo: Sandy as Bad as Anything I've Experienced in New York," *Daily News*, October 30, 2012; http://www.nydailynews.com/blogs/daily politics/gov-cuomo-sandy-bad-experienced-new-york-blog-entry-1.1692632.

2. M. Sledge, "Hurricane Sandy Shows We Need to Prepare for Climate Change, Cuomo and Bloomberg Say," *Huffington Post*, October 30, 2012. http://www .huffingtonpost.com/2012/10/30/hurricane-sandy-cuomo-bloomberg-climate -change_n_2043982.html.

3. Bronx River Alliance. "The Bronx River Current—Hurricane Sandy Issue," November 2012, http://bronxriver.org/puma/images/usersubmitted/file/November %202012.pdf.

4. A. Revkin, "Students Press the Case for Oysters as New York's Surge Protector," *New York Times*, November 12, 2012. http://dotearth.blogs.nytimes .com/2012/11/12/students-press-the-case-for-oysters-as-new-yorks-surge -protector/.

5. M. Schwartz, "On Decimated Shore, a Second Life for Christmas Trees," *New York Times*, February 3, 2013, http://www.nytimes.com/2013/02/04/nyregion/ on-decimated-shore-a-second-life-for-christmas-trees.html?_r=0. Quotes in the article from Long Beach residents lend support to our claim throughout this book that post-disaster stewardship efforts contribute to healing and to bringing the community together:

> "It was a very nice healing thing for residents to do to contribute to our protection," said Jack Schnirman, the city manager.
>
> The town and local volunteers have promoted several other projects to aid the recovery while also lifting spirits. There have been benefit cookouts and street festivals to help local businesses. Residents have planted what city officials called "recovery bulbs," which will bloom into flowers in spring.
>
> "There are so many things bringing this community together, which is great," said Alison Kallelis, 33, who was among those to propose the Christmas tree idea.
>
> If all goes well, Ms. Kallelis said, laying the trees on the beach could become an annual tradition here.
>
> "Every year you keep adding more trees," she said, "and keep building it up higher and higher."

6. P. Cafaro, "Thoreau, Leopold, and Carson: Toward an Environmental Virtue Ethics," *Environmental Ethics* 22, no. 4 (2001): 3–17.

7. A. Næss, "The Deep Ecology 'Eight Points' Revisited," in *The Selected Works of Arne Næss*, ed. Harold Glasser and Alan Drengson (Dordrecht, The Netherlands: Springer, 2005), 57–66.

# Resources

Aguilar, A. "Kane County Veteran Helps Younger Vets Find Peace through Nature." *Chicago Tribune*, October 25, 2012.

Ahn, T. K., and E. Ostrom, eds. *Foundations of Social Capital*. Chelthenham, UK: Edward Elgar Publishing, 2003.

Ahn, T. K., and E. Ostrom. "Social Capital and Collective Action." In *The Handbook of Social Capital*, ed. D. Castiglione, J. W. van Deth, and G. Wolleb, 70–100. Oxford: Oxford University Press, 2008.

Albrecht, G., G.-M. Sartore, L. Connor, N. Higginbotham, S. Freeman, B. Kelly, H. Stain, A. Tonna, and G. Pollard. "Solastalgia: The Distress Caused by Environmental Change." *Australasian Psychiatry* 15 (2007): S95–S98.

Alfaro, Y. "The Bronx River Alliance." *The Greener Apple*, December 18, 2009. http://macaulay.cuny.edu/eportfolios/thegreenerapple/2009/12/18/the-bronx-river-alliance/, accessed March 28, 2012.

Allen, E., B. Beckwith, J. Beckwith, S. Chorover, D. Culver, M. Duncan, S. Gould, et al. "Against 'Sociobiology.'" *New York Review of Books* 182 (1975): 184–186. http://www.nybooks.com/articles/archives/1975/nov/13/against-sociobiology/, accessed July 7, 2014.

Allweil, Y. "Shrinking Cities: Like a Slow-Motion Katrina." *Forum of Design for the Public Realm* 19, no. 1 (2007): 91–93. http://places.designobserver.com/media/pdf/Dispatches__--_456.pdf, accessed January 24, 2014.

Amsden, B. L., R. C. Stedman, and L. E. Kruger. "The Creation and Maintenance of Sense of Place in a Tourism-Dependent Community." *Leisure Sciences* 33, no. 1 (2010): 32–51.

Andersson, E., S. Barthel, and K. Ahrne. "Measuring Social-Ecological Dynamics behind the Generation of Ecosystem Services." *Ecological Applications* 17, no. 5 (2007): 1267–1278.

Anon. "Conservationist Found Murdered in Cape Town Home." *Grassroots: Newsletter of the Grassland Society of Southern Africa* 8, no. 1 (February 2008): 2. http://grassland.org.za/resources/publications/grassroots/2008/February%202008/1%20Regulars%20Feb%202008.pdf/view, accessed July 6, 2014.

Ansari, S. "Social Capital and Collective Efficacy: Resource and Operating Tools of Community Social Control." *Journal of Theoretical and Philosophical Criminology* 5, no. 2 (2013): 75–94.

Armitage, D., F. Berkes, and N. Doubleday. "Introduction: Moving Beyond Co-management." In *Adaptive Co-management: Collaboration, Learning, and Multi-level Governance*, ed. D. Armitage, F. Berkes, and N. Doubleday, 1–18. Vancouver: UBC Press, 2007.

Armitage, D., M. Marschke, and R. Plummer. "Adaptive Co-Management and the Paradox of Learning." *Global Environmental Change* 18 (2008): 86–98.

Ashden Awards. "Trees, Water and People (TWP) and Asociación Hondureña Para El Desarollo (HDESA), Honduras." December 2009. http://www.ashden.org /files/TWP%20full.pdf, accessed July 6, 2014.

Ashoka. "Mandla Mentoor." N.d. https://www.ashoka.org/fellow/mandla-mentoor, accessed July 7, 2014.

Atkinson, J. "An Evaluation of the Gardening Leave Project for Ex-Military Personnel with PTSD and Other Combat-Related Mental Health Problems." Glasgow, UK: The Pears Foundation, 2009.

Austin, M. E., and R. Kaplan. "Identity, Involvement, and Expertise in the Inner City: Some Benefits of Tree-Planting Projects." In *Identity and the Natural Environment: The Psychological Significance of Nature*, ed. S. Clayton and S. Opotow, 205–225. Cambridge, MA: MIT Press, 2003.

Bandura, A. "Perceived Self-Efficacy in Cognitive Development and Functioning." *Educational Psychologist* 28, no. 2 (1993): 117–148.

Bandura, A. "Self-Efficacy: Toward a Unifying Theory of Behavioral Change." *Psychological Review* 84, no. 2 (1977): 191–215.

Bandura, A. *Social Learning Theory*. Englewood Hills, NJ: Prentice-Hall, Inc., 1977.

Barab, S. A., and W.-M. Roth. "Curriculum-Based Ecosystems: Supporting Knowing from an Ecological Perspective." *Educational Researcher* 35, no. 5 (2006): 3–13.

Barthel, S., J. Colding, T. Elmqvist, and C. Folke. "History and Local Management of a Biodiversity-Rich, Urban Cultural Landscape." *Ecology and Society* 10, no. 2 (2005), http://www.ecologyandsociety.org/vol10/iss2/art10/, accessed July 3, 2014.

Barthel, S., C. Folke, and J. Colding. "Social-Ecological Memory in Urban Gardens: Retaining the Capacity for Management of Ecosystem Services." *Global Environmental Change* 20 (2010): 255–265.

Barthel, S., J. Parker, C. Folke, and J. Colding. "Urban Gardens: Pockets of Social-Ecological Memory." In *Greening in the Red Zone*, ed. K. G. Tidball and M. E. Krasny, 145–158. New York: Springer, 2014.

Beatley, T. *Biophilic Cities: Integrating Nature into Urban Design and Planning*. Washington, DC: Island Press, 2011.

Becker, V. J. "Regreening Instead of Guarding," *Neues Deutschland Sozialistische Tageszeitung*, trans. Bernd Blossey, August 14, 2010, http://www.neues-deutschland .de/artikel/177381.begruenen-statt-bewachen.html, accessed July 7, 2014.

Benight, C. C., and A. Bandura. "Social Cognitive Theory of Posttraumatic Recovery: The Role of Perceived Self-Efficacy." *Behaviour Research and Therapy* 42 (2004): 1129–1148.

Bergman, M. "Athlone History: The Trojan Horse Massacre." N.d. http://www .athlone.co.za/heritage/history.php, accessed July 6, 2014.

Berkes, F. "Knowledge, Learning and the Resilience of Social-Ecological Systems." In *Knowledge for the Development of Adaptive Co-Management*. Oaxaca, MX: Tenth Biennial Conference of the International Association for the Study of Common Property, 2004. http://dlc.dlib.indiana.edu/dlc/bitstream/handle/10535/2385 /berkes_knowledge_040511_paper299a.pdf?sequence=1, accessed July 3, 2014.

Berkes, F., J. Colding, and C. Folke. *Navigating Social-Ecological Systems: Building Resilience for Complexity and Change*. Cambridge, UK: Cambridge University Press, 2003.

Berkes, F., J. Colding, and C. Folke. "Rediscovery of Traditional Ecological Knowledge as Adaptive Management." *Ecological Applications* 10, no. 5 (2000): 1251–1262.

Berkes, F., and C. Folke. "Back to the Future: Ecosystem Dynamics and Local Knowledge." In *Panarchy: Understanding Transformations in Human and Natural Systems*, ed. L. H. Gunderson and C. S. Holling, 121–146. Washington, DC: Island Press, 2002.

Berkes, F., and C. Folke. *Linking Social and Ecological Systems*. London: Cambridge University Press, 1998.

Berkes, F., and N. J. Turner. "Knowledge, Learning and the Evolution of Conservation Practice for Social-Ecological System Resilience." *Human Ecology* 34, no. 4 (2006): 479–494.

Bernstein, F. A. "A Vision of a Park on a Restored Los Angeles River." *New York Times*, September 28, 2010. http://www.nytimes.com/2010/09/29 /realestate/29river.html?_r=1, accessed July 5, 2014.

Biggs, R., F. Westley, and S. Carpenter. "Navigating the Back Loop: Fostering Social Innovation and Transformation in Ecosystem Management." *Ecology and Society* 15, no. 2 (2010). http://www.ecologyandsociety.org/vol15/iss2/art9/, accessed July 5, 2014.

Blackmore, C. "What Kinds of Knowledge, Knowing and Learning Are Required for Addressing Resource Dilemmas?: A Theoretical Overview." *Environmental Science & Policy* 10 (2007): 512–525.

Blackmore, C., R. Ison, and J. Jiggins. "Social Learning: An Alternative Policy Instrument for Managing in the Context of Europe's Water." *Environmental Science & Policy* 10, no. 6 (2007): 493–498.

Bodin, O., and B. I. Crona. "The Role of Social Networks in Natural Resource Governance: What Relational Patterns Make a Difference?" *Global Environmental Change* 19 (2009): 366–374.

Bolund, P., and S. Hunhammar. "Ecosystem Services in Urban Areas." *Ecological Economics* 29 (1999): 293–301.

Bonanno, G. A. "Loss, Trauma, and Human Resilience: How We Have Underestimated the Human Capacity to Thrive after Extremely Aversive Events." *American Psychologist* 59, no. 1 (2004): 20–28.

Bonney, R., C. B. Cooper, J. Dickinson, S. Kelling, T. Phillips, K. V. Rosenberg, and J. Shirk. "Citizen Science: A Developing Tool for Expanding Science Knowledge and Scientific Literacy." *BioScience* 59, no. 11 (2009): 977–984.

Bookchin, M. "What Is Social Ecology?" In *Environmental Philosophy: from Animal Rights to Radical Ecology*, ed. M. E. Zimmerman, J. B. Callicott, G. Sessions, K. J. Warren, and J. Clark, 354–373. Englewood Cliffs, NJ: Prentice Hall, 1993.

Borden, R. J. "A Brief History of SHE: Reflections on the Founding and First Twenty Five Years of the Society for Human Ecology." *Human Ecology Review* 15, no. 1 (2008): 95–108.

Boukharaeva, L. "Six Ares of Land for the Resilience of Urban Families in Post-Soviet Russia." In *Greening in the Red Zone*, ed. K. G. Tidball and M. E. Krasny, 357–360. New York: Springer, 2014.

Bowles, S., S. N. Durlauf, and K. Hoff. *Poverty Traps*. Princeton: Princeton University Press, 2006.

Branas, C. C., R. A. Cheney, J. M. MacDonald, V. W. Tam, T. D. Jackson, and T. R. Ten-Have. "A Difference-in-Differences Analysis of Health, Safety, and Greening Vacant Urban Space." *American Journal of Epidemiology* 174, no. 11 (2011): 1296–1306.

Brenner, E. "Restored River a Boon to Yonkers." *New York Times*, August 9, 2012. http://www.nytimes.com/2012/08/12/realestate/westchester-in-the-region-restored-river-a-boon-to-yonkers.html, accessed July 7, 2014.

Briggs, L., and M. E. Krasny. "Conservation in Cities: Linking Citizen Science and Civic Ecology Practices." Ithaca, NY: Civic Ecology Lab, Cornell University, 2013. http://civeco.files.wordpress.com/2013/09/2013-briggs.pdf, accessed July 7, 2014.

Bronx River Alliance. "History." N.d. http://bronxriver.org/?pg=content&p=aboutus&m1=44, accessed July 7, 2014.

Bronx River Alliance. "The Bronx River Current—Hurricane Sandy Issue," November 2012. http://bronxriver.org/puma/images/usersubmitted/file/November%202012.pdf, accessed July 9, 2014.

Bullard, R. D., and G. S. Johnson. "Environmental Justice: Grassroots Activism and Its Impact on Public Policy Decision Making." *Journal of Social Issues* 56, no. 3 (2000): 555–578.

Buranen, M. "Reforest the Bluegrass." Stormwater, March 16, 2000. http://www.stormh2o.com/SW/articles/16503.aspx, accessed July 7, 2014.

Burch, W., and J. M. Grove. "Ecosystem Management: Some Social, Conceptual, Scientific, and Operational Guidelines for Practioners." In *Ecological Stewardship: A Common Reference for Ecosystem Management*, ed. N. C. Johnson, A.

J. Malk, W. T. Sexton, and R. C. Szaro, 279–295. Oxford: Elsevier Science Ltd, 1999.

Cafaro, P. "Thoreau, Leopold, and Carson: Toward an Environmental Virtue Ethics." *Environmental Ethics* 22, no.4 (2001): 3–17.

Campbell, L. "Civil Society Strategies on Urban Waterways: Stewardship, Contention, and Coalition Building." Master's thesis, Department of Urban Studies and Planning, Massachusetts Institute of Technology, 2006.

Carpenter, S. R., and W. A. Brock. "Adaptive Capacity and Traps." *Ecology and Society* 13, no. 2 (2008), http://www.ecologyandsociety.org/vol13/iss2/art40/, accessed July 6, 2014.

Carpenter, S. R., D. Ludwig, and W. A. Brock. "Management of Eutrophication for Lakes Subject to Potentially Irreversible Change." *Ecological Applications* 9, no. 3 (1999): 751–771.

Castiglione, D. "Social Capital as a Research Programme." In The *Handbook of Social Capital*, ed. D. Castiglione, J. W. van Deth, and G. Wolleb, 177–195. Oxford: Oxford University Press, 2008.

Castiglione, D., J. W. van Deth, and G. Wolleb, eds. *The Handbook of Social Capital*. Oxford: Oxford University Press, 2008.

CEC. "Citizens' Environmental Coalition." http://www.cechouston.org/, accessed July 7, 2014.

Center for Biological Diversity. "Bald Eagle Population Exceeds 11,000 Pairs in 2007." N.d. http://www.biologicaldiversity.org/species/birds/bald_eagle/report/, accessed January 29, 2014.

Central Park Conservancy. "Official Website of New York City's Central Park." N.d. http://www.centralparknyc.org/about/, accessed January 29, 2014.

Chan, K. M. A., A. D. Guerry, P. Balvanera, S. Klain, T. Satterfield, X. Basurto, A. Bostrom, et al. "Where Are Cultural and Social in Ecosystem Services? A Framework for Constructive Engagement." *BioScience* 62, no. 8 (2012): 744–756.

Chan, K. M. A., T. Satterfield, and J. Goldstein. "Rethinking Ecosystem Servcies to Better Address and Navigate Cultural Values." *Ecological Economics* 74 (2012): 8–18.

Chapin, F. S., M. E. Power, S. T. A. Pickett, A. Freitag, J. A. Reynolds, R. B. Jackson, D. M. Lodge, C. Duke, S. L. Collins, A. G. Power, and A. Bartuska. "Earth Stewardship: Science for Action to Sustain the Human-Earth System." *Ecosphere* 2, no. 8 (2011): http://www.esajournals.org/doi/full/10.1890/ES11-00166.1, accessed July 7, 2014.

Chawla, L. "Children's Engagement with the Natural World as a Ground for Healing." In *Greening in the Red Zone*, ed. K. G. Tidball and M. E. Krasny, 111–124. New York: Springer, 2014.

Chawla, L. "Learning to Love the Natural World Enough to Protect It." *Barn* 2 (2006): 57–78.

Chawla, L. "Participation and the Ecology of Environmental Awareness." In *Participation and Learning: Perspectives on Education and the Environment, Health and Sustainability*, ed. A. Reid, B. B. Jensen, and J. Nikel, 98–110. New York: Springer-Verlag, 2008.

Cheng, S., and J. R. McBride. "Restoration of the Urban Forests of Tokyo and Hiroshima Following World War II." In *Greening in the Red Zone*, ed. K. G. Tidball and M. E. Krasny, 225–248. New York: Springer, 2014.

City of Houston. "Green Houston." 2014. http://www.greenhoustontx.gov/, accessed January 29, 2014.

City of Houston. "Houston Community Gardening Program." 2014. http://www.houstontx.gov/health/Community/garden.html, accessed January 29, 2014.

Colding, J., S. Barthel, P. Bendt, R. Snep, W. van der Knapp, and H. Ernstson. "Urban Green Commons: Insights on Urban Common Property Systems." *Global Environmental Change* 23 (2013): 1039–1051.

Coleman, J. S. "Social Capital in the Creation of Human Capital." *American Journal of Sociology* 94 (Supplement) (1988): S95–S120.

Compost for Brooklyn. N.d. http://compostforbrooklyn.org/about/, accessed July 3, 2014.

Connolly, J. J., E. S. Svendsen, D. R. Fisher, and L. K. Campbell. "Organizing Urban Ecosystem Services through Environmental Stewardship Governance in New York City." *Landscape and Urban Planning* 109 (2013): 76–84.

Cornell Cooperative Extension. "New York Invasive Species Clearinghouse." 2014. http://www.nyis.info/index.php?action=eab, accessed July 7, 2014.

Cornwall, A., and R. Jewkes. *What Is Participatory Research? Social Science & Medicine* 41, no. 12 (1995): 1667–1676.

Costanza, R., R. d'Arge, R. de Groot, S. Farber, M. Grasso, B. Hannon, K. Limburg, et al. "The Value of the World's Ecosystem Services and Natural Capital." *Nature* 387 (1997): 253–260.

Cramer, M. "The Berlin Wall Trail: A Cycling and Hiking Route on the Traces of Berlin's East-West Division during the Cold War." In *Greening in the Red Zone*, ed. K. G. Tidball and M. E. Krasny, 445–449. New York: Springer, 2014.

Davidson-Hunt, I., and F. Berkes. "Learning as You Journey: Anishinaabe Perception of Social-Ecological Environments and Adaptive Learning." *Conservation Ecology* 8, no. 1 (2003). http://www.consecol.org/vol8/iss1/art5/, accessed July 3, 2014.

Davidson-Hunt, I., and M. O'Flaherty. "Researchers, Indigenous Peoples, and Place-Based Learning Communities." *Society & Natural Resources* 20 (2007): 291–305.

Delia, J. E. "Cultivating a Culture of Authentic Care in Urban Environmental Education: Narratives from Youth Interns at East New York Farms!" M.S. thesis. Ithaca, NY: Cornell University, 2013.

Díaz, E., N. Bruce, D. Pope, A. Díaz, K. Smith, and T. Smith-Sivertsen. "Self-Rated Health among Mayan Women Participating in a Randomised Intervention Trial Reducing Indoor Air Pollution in Guatemala." *BMC International Health and Human Rights* 8, no.1 (2008): 1–8.

Díaz, S., F. Quétier, D. M. Cáceres, S. F. Trainor, N. Pérez-Harguindeguy, M. S. Bret-Harte, B. Finegan, M. Peña-Claros, and L. Poorterg. "Linking Functional Diversity and Social Actor Strategies in a Framework for Interdisciplinary Analysis of Nature's Benefits to Society." *Proceedings of the National Academy of Sciences of the United States of America* 108, no. 3 (2011): 895–902.

Dickinson, J. L., and R. Bonney, eds. *Citizen Science: Public Collaboration in Environmental Research*. Ithaca, NY: Cornell University Press, 2012.

Dictionary.com. "Oneg Shabbat." 2014. http://dictionary.reference.com/browse /oneg+shabbat, accessed July 7, 2014

Dietz, T., E. Ostrom, and P.C. Stern. "The Struggle to Govern the Commons." *Science* 302, no. 5652 (December 12, 2003): 1907–1912.

Donovan, G. H., D. T. Butry, Y. L. Michael, J. P. Prestemon, A. M. Liebhold, D. Gatziolis, and M. Y. Mao. "The Relationship between Trees and Human Health: Evidence from the Spread of the Emerald Ash Borer." *American Journal of Preventive Medicine* 44, no. 2 (2013): 139–145.

DuBois, B., and K. G. Tidball. "Greening in Coastal New York Post-Sandy." Civic Ecology Lab White Paper Series. #CEL004. Ithaca, NY: Cornell Civic Ecology Lab, Cornell University, 2013. http://civeco.files.wordpress.com/2013/09/2013 -dubois.pdf, accessed August 1, 2014.

Dwyer, J. "New Daffodils for Garden That Outlived a Creator." *New York Times*, November 30, 2001. http://www.nytimes.com/2001/11/30/nyregion/a-nation -challenged-objects-new-daffodils-for-garden-that-outlived-a-creator.html, accessed January 24, 2014.

Earls, F., and M. Carlson. "The Social Ecology of Child Health and Well-Being." *Annual Review of Public Health* 22 (2001): 143–166.

EarthCorps. "Local Restoration, Global Leadership: EarthCorps." N.d. http:// www.earthcorps.org/index.php, accessed July 3, 2014.

Ehrlich, P. R., and H. A. Mooney. "Extinction, Substitution, and Ecosystem Services." *BioScience* 33, no. 4 (1983): 248–254.

Eisen, G. *Children and Play in the Holocaust: Games among the Shadows*. Amherst: University of Massachusetts Press, 1988.

Eisenman, T. S. "Frederick Law Olmsted, Green Infrastructure, and the Evolving City." *Journal of Planning History* 12 (2013): 287–311.

Elmqvist, T., M. Fragkias, J. Goodness, B. Güneralp, P. Marcotullio, R. McDonald, S. Parnell, et al. *Urbanization, Biodiversity and Ecosystem Services: Challenges and Opportunities*. New York: Springer, 2013.

Encyclopedia Brittanica. "Oneg Shabbat." February 19, 2014. http://www.britannica .com/EBchecked/topic/429113/Oneg-Shabbat, accessed July 7, 2014.

Engeström, Y. "Expansive Learning at Work: Toward an Activity-Theoretical Conceptualization." *Journal of Education and Work* 14, no. 1 (2001): 133–156.

Engeström, Y. *Learning by Expanding: An Activity-Theoretical Approach to Developmental Research*. Helsinki: Orienta-Konsultit, 1987.

EPA. "Urban Waters Federal Partnership." June 13, 2014. http://www.urbanwaters .gov/, accessed July 7, 2014.

Erikson, E. H., J. M. Erikson, and H. Q. Kivnick. *Vital Involvement in Old Age.* New York: W. W. Norton & Company, 1986.

Ernstson, H., S. Barthel, E. Andersson, and S. T. Borgström. "Scale-Crossing Brokers and Network Governance of Urban Ecosystem Services: The Case of Stockholm." *Ecology and Society* 15, no. 4 (2010), http://www.ecologyandsociety.org /vol15/iss4/art28/, accessed July 7, 2014.

Ernstson, H., and T. Elmqvist. "Globalization, Urban Ecosystems and Social-Ecological Innovations: A Comparative Network Analytic Approach." Stockholm: Stockholm University, 2011.

Eternal Reefs Inc. "Eternal Reefs." 2014. http://www.eternalreefs.com/, accessed January 29, 2014.

Evelly, J. "Bronx Group Turns Parking Space into Temporary Public 'Park.'" *DNAinfo New York*, September 26, 2012. http://www.dnainfo.com/new-york /20120926/soundview/bronx-group-turns-parking-space-into-temporary-public -park, accessed July 7, 2014.

Faber, D., and D. McCarthy. "The Evolving Structure of the Environmental Justice Movement in the United States: New Models for Democratic Decision-Making." *Social Justice Research* 14, no. 4 (2001): 405–421.

Faber Taylor, A., and F. E. Kuo. "Children with Attention Deficits Concentrate Better after Walk in the Park." *Journal of Attention Disorders* 12 (2009): 402–409.

Faber Taylor, A., F. E. Kuo, and W. C. Sullivan. "Coping with ADD: The Surprising Connection to Green Play Settings." *Environment and Behavior* 33, no. 1 (2001): 54–77.

Faber Taylor, A., A. Wiley, F. E. Kuo, and W. C. Sullivan. "Growing up in the Inner City: Green Spaces as Places to Grow." *Environment and Behavior* 30, no. 1 (1998): 3–27.

Fahrenthold, D. A. "Last House on Sinking Chesapeake Bay Island Collapses." *Washington Post*, October 26, 2010. http://www.washingtonpost.com/wp-dyn/ content/article/2010/10/24/AR2010102402996.html, accessed July 5, 2014.

Farlex. *The Free Dictionary*. N.d. http://www.thefreedictionary.com/vicious+cycle, accessed January 29, 2014.

Feedback Farms. http://www.feedbackfarms.com/, accessed July 14, 2014.

Fisher, D. R., L. Campbell, and E. S. Svendsen. "The Organisational Structure of Urban Environmental Stewardship." *Environmental Politics* 21, no. 1 (2012): 26–48.

Fligstein, N. "Institutional Entrepreneurs and Cultural Frames—the Case of the European Union's Single Market Program." *European Societies* 3, no. 3 (2001): 261–287.

FoLAR. "Friends of the Los Angeles River." N.d. http://folar.org/, accessed July 7, 2014.

Folke, C. "Resilience: The Emergence of a Perspective for Social-Ecological Systems Analyses." *Global Environmental Change* 16 (2006): 253–267.

Folke, C., S. R. Carpenter, B. Walker, M. Scheffer, T. Chapin, and J. Rockström. "Resilience Thinking: Integrating Resilience, Adaptability and Transformability." *Ecology and Society* 15, no. 4 (2010), http://www.ecologyandsociety.org/vol15/iss4/art20/, accessed November 22, 2010.

Folke, C., S. Carpenter, T. Elmqvist, L. Gunderson, C. S. Holling, B. Walker, J. Bengtsson, et al. "Resilience and Sustainable Development: Building Adaptive Capacity in a World of Transformations." Johannesburg, South Africa: World Summit on Sustainable Development, 2002. http://www.resalliance.org/files/1144440669_resilience_and_sustainable_development.pdf, accessed July 3, 2014.

Forterra. "Forterra." N.d. http://www.forterra.org/, accessed July 7, 2014.

Francis, M., L. Cashdan, and L. Paxson. *Community Open Spaces*. Washington, DC: Island Press, 1984.

Friedland, L. A., and S. Morimoto. "The Changing Lifeworld of Young People: Risk, Resume-Padding, and Civic Engagement." The Center for Information & Research on Civic Learning and Engagement (CIRCLE). College Park, MD: CIRCLE, 2005.

Friedman, S. S. *Amcha: An Oral Testament of the Holocaust*. Washington, DC: University Press of America, 1979.

Friends of Mitchell Park. N.d. www.mitchellparkdc.org/, accessed August 1, 2014.

Friends of the Rouge Watershed. N.d. http://www.frw.ca/, accessed July 7, 2014.

Fromm, E. *The Heart of Man*. New York: Harper & Row, 1964.

Gadgil, M., N. S. Hemam, and B. M. Reddy. "People, Refugia and Resilience." In *Linking Social and Ecological Systems: Management Practices and Social Mechanisms for Building Resilience*, ed. F. Berkes and C. Folke, 30–47. Cambridge, UK: Cambridge University Press, 1998.

Gadgil, M., P. Olsson, F. Berkes, and C. Folke. "Exploring the Role of Local Ecological Knowledge in Ecosystem Management: Three Case Studies." In *Navigating Social-Ecological Systems: Building Resilience for Complexity and Change*, ed. F. Berkes, J. Colding, and C. Folke, 189–209. Cambridge, UK: Cambridge University Press, 2003.

Geisler, C. "Green Zones from Above and Below: A Retrospective Cautionary Tale." In *Greening in the Red Zone*, ed. K. G. Tidball and M. E. Krasny, 203–215. New York: Springer, 2014.

Gibson, J. J. "The Theory of Affordances." In *Perceiving, Acting, and Knowing: Toward an Ecological Psychology*, ed. Robert Shaw and John Bransford, 67–82. Hillsdale: Lawrence Erlbaum Associates, 1977.

Gies, E. "Restoring Iraq's Garden of Eden." *New York Times*, April 17, 2013. http://www.nytimes.com/2013/04/18/world/middleeast/restoring-iraqs-garden-of-eden.html?pagewanted=all&_r=0, accessed August 1, 2014.

Gladwell, M. *The Tipping Point: How Little Things Can Make a Big Difference.* New York: Little, Brown and Company, 2002.

Golshani, Z., and M. E. Krasny. *Civic Ecology Practice in Iran: The Case of the Nature Cleaners.* Baltimore, MD: North American Association for Environmental Education, 2013.

Gómez-Baggethun, E., R. de Groot, P. L. Lomas, and C. Montes. "The History of Ecosystem Services in Economic Theory and Practice: From Early Notions to Markets and Payment Schemes." *Ecological Economics* 69 (2010): 1209–1218.

Goncalves, M. "Environmentalists Celebrate Return of Alewife." *The Hunts Point Express,* May 28, 2009, http://brie.hunter.cuny.edu/hpe/2009/05/28/little-fish -signal-big-changes-in-bronx-river/, accessed July 7, 2014.

Gonzalez, E. D. *The Bronx.* New York: Columbia University Press, 2004.

Gottlieb, R. *Reinventing Los Angeles.* Cambridge, MA: MIT Press, 2007.

Gottlieb, R., and A. Azuma. "Re-envisioning the Los Angeles River: An NGO and Academic Institute Influence the Policy Discourse." *Golden Gate University Law Review* 35, no. 3 (2005): 321–342.

Gough, S., and W. Scott. "Promoting Environmental Citizenship through Learning: Toward a Theory of Change." In *Environmental Citizenship,* ed. A. Dobson and D. Bell, 263–285. Cambridge, MA: MIT Press, 2006.

Gowanus Canal Conservancy. N.d. http://www.gowanuscanalconservancy.org/ ee/, accessed September 15, 2014.

"Gowanus Canal Conservancy Blog." June 27, 2014. http://gowanuscanalconser vancy.wordpress.com/, accessed July 7, 2014.

Grant, S., and M. Humphries. "Critical Evaluation of Appreciative Inquiry: Bridging an Apparent Paradox." *Action Research* 4, no. 4 (2006): 401–418.

Great Sunflower Project. "The Great Sunflower Project." N.d. http://www.great sunflower.org/, accessed July 7, 2014.

Green, A. "South Bronx Nonprofit Rocking the Boat Teams with Ecological Artist Lillian Ball, Edesign Dynamics, Drexel University and ABC Carpet and Home on Project to Reduce Pollution in Bronx River." October 25, 2011. http://long islandsoundstudy.net/wp-content/uploads/2011/10/WATERWASH-ABC-Media -Advisory.pdf, accessed July 7, 2014.

Greeno, J. G. "Gibson's Affordances." *Psychological Review* 101, no. 2 (1994): 336–342.

Greeno, J. G. "Situativity of Knowing, Learning, and Research." *American Psychologist* 53, no. 1 (1998): 5–26.

Gregory, M. "The Garden Ecology Project." January 2011. blogs.cornell.edu/gep /files/2011/02/GEP-Gardener-Update.pdf, accessed July 7, 2014.

Grichting, A. "Cyprus: Greening in the Dead Zone." In *Greening in the Red Zone,* ed. K. G. Tidball and M. E. Krasny, 429–443. New York: Springer, 2014.

Grichting, A., and K. G. Kim. "The Korea DMZ: From a Red Zone to a Deeper Shade of Green." In *Greening in the Red Zone,* ed. K. G. Tidball and M. E. Krasny, 197–201. New York: Springer, 2014.

Groundwork Hudson Valley. "Saw Mill River Coalition." N.d. http://www .sawmillrivercoalition.org/, accessed July 7, 2014.

Grove, J. M. "Cities: Managing Densely Settled Social-Ecological Systems." In *Principles of Ecosystem Stewardship. Resilience-Based Natural Resource Management in a Changing World*, ed. F. S. Chapin, G. P. Kofinas, and C. Folke, 281–294. New York: Springer, 2009.

Gunderson, L. H., and C. S. Holling, eds. *Panarchy: Understanding Transformations in Human and Natural Systems*. Washington, DC: Island Press, 2002.

Hanifan, L. J. "The Rural School Community Center." *Annals of the American Academy of Political and Social Science* 67 (1916): 130–138.

Hansen, J., P. Kharecha, M. Sato, V. Masson-Delmotte, F. Ackerman, D. J. Beerling, P. J. Hearty, et al. "Assessing 'Dangerous Climate Change': Required Reduction of Carbon Emissions to Protect Young People, Future Generations and Nature." PLoS ONE 8, no. 12 (2013): e81648. http://dx.doi.org/10.1371%2Fjournal .pone.0081648, accessed August 2, 2014.

Hansen, L. "'History on the 'Half Shell' in 'Big Oyster.'" National Public Radio, April 6, 2006, http://www.npr.org/templates/story/story.php?storyId=5332982, accessed July 3, 2014.

Hardin, G. "The Tragedy of the Commons." *Science* 162, no. 3859 (1968): 1243–1248.

Harper, D. "Online Etymology Dictionary." 2014. http://www.etymonline.com/, accessed July 3, 2014.

Harris, L. E. "Artificial Reefs for Ecosystem Restoration and Coastal Erosion Protection with Aquaculture and Recreational Amenities." 2006. http://www.reefball .org/album/%3D%3D%29%20Non-Geographic%20defined%20Photos/arti ficialreefscientificpapers/2006JulyLEHRBpaper.pdf, accessed January 29, 2014.

Hartig, T., M. Mang, and G. W. Evans. "Restorative Effects of Natural Environment Experiences." *Environment and Behavior* 23, no. 1 (1991): 3–26.

Heberlein, T. A. *Navigating Environmental Values*. Oxford: Oxford University Press, 2012.

Heikkila, T., and A. K. Gerlak. "The Formation of Large-Scale Collaborative Resource Management Institutions: Clarifying the Roles of Stakeholders, Science, and Institutions." *Policy Studies Journal: The Journal of the Policy Studies Organization* 33, no. 4 (2005): 583–612.

Helphand, K. *Defiant Gardens: Making Gardens in Wartime*. San Antonio, TX: Trinity University Press, 2006.

Hike for KaTREEna. N.d. http://www.hikeforkatreena.com/, accessed July 7, 2014.

History com. "Berlin Wall." N.d. http://www.history.com/topics/cold-war/berlin -wall, accessed July 7, 2014.

Holling, C. S. "The Spruce-Budworm/Forest-Management Problem." In *Adaptive Environmental Assessment and Management*, ed. C. S. Holling, 143–182. Hoboken, NJ: John Wiley & Sons, 1978.

258    Resources

Holling, C. S., L. H. Gunderson, and G. D. Peterson. "Sustainability and Panarchies." In *Panarchy: Understanding Transformations in Human and Natural Systems*, ed. L. H. Gunderson and C. S. Holling, 63–102. Washington, DC: Island Press, 2002.

Hopkins, A. W. *Groundswell: Stories of Saving Places, Finding Community*. San Francisco: The Trust for Public Land, 2005.

Hopkins, G. M. *Poems and Prose*, ed. W. H. Gardner. London: Penguin Classics, 1985.

Houston Parks and Recreation Department. "Current Community Gardens." 2014. http://www.houstontx.gov/parks/urbangardener/urbangardener_Current .html, accessed July 3, 2014.

Huber, K. "Alabama Grows More Than Veggies: It Grows Spirit." *Houston Chronicle*, March 10, 2007. http://www.chron.com/life/gardening/article/Alabama -grows-more-than-veggies-It-grows-spirit-1832754.php, accessed July 7, 2014.

Hudson, C. "Community Hero: J. A. and Geraldine Reynolds." August 27, 2011. http://myhero.com/go/hero.asp?hero=Bruce_Garden_Sept11, accessed July 7, 2014.

Hughes, J. D. "Pan: Environmental Ethics in Classical Polytheism." In *Religion and Environmental Crisis*, ed. E. C. Hargrove, 7–24. Athens: University of Georgia Press, 1986.

Huitema, D., E. Mostert, W. Egas, S. Moellenkamp, C. Pahl-Wostl, and R. Yalcin. "Adaptive Water Governance: Assessing the Institutional Prescriptions of Adaptive (Co-)Management from a Governance Perspective and Defining a Research Agenda." *Ecology and Society* 14, no. 1 (2009), http://www.ecologyandsociety .org/vol14/iss1/art26/, accessed July 7, 2014.

Hurley, D. "Scientist at Work—Felton Earls; On Crime as Science (a Neighbor at a Time)." *New York Times*, January 6, 2004. http://www.nytimes.com/2004/01/06 /science/scientist-at-work-felton-earls-on-crime-as-science-a-neighbor-at-a-time .html, accessed July 7, 2014.

Hwang, H., and W. J. Powell. "Institutions and Entrepreneurship." In *The Handbook of Entrepreneurship*, ed. S. A. Alvarez, R. Agrawal, and O. Sorenson, 210–232. New York: Springer, 2005.

ICLEI. "FAQ: ICLEI, the United Nations, and Agenda 21." N.d. http://www .icleiusa.org/about-iclei/faqs/faq-iclei-the-united-nations-and-agenda-21#what -is-agenda-21, accessed July 7, 2014.

ICLEI. "URBIS—Urban Biosphere Initiative." 2014. http://urbis.iclei.org/, accessed January 29, 2014.

Jackson, L. P. "Storm Fuels Environmental Awareness: A Guest Column by Lisa P. Jackson." *The Times-Picayune*, August 26, 2010. http://www.nola.com/katrina/ index.ssf/2010/08/storm_fuels_environmental_awar.html, accessed July 6, 2014.

Jerusalem Municipality. "The Gazelle Valley Urban Nature Park." N.d. http:// greenpilgrimjerusalem.org/symposium2013/wp-content/uploads/sites/3/2013/02 /Gazelle-Valley-Intro.pdf, accessed July 7, 2014.

John, D. "Civic Environmentalism." In *Environmental Governance Reconsidered*, ed. R. F. Durant, D. Fiorini, and R. O'Leary, 219–254. Cambridge, MA: MIT Press, 2004.

Kahn, P. H., Jr., B. Friedman, B. Gill, J. Hagman, R. L. Severson, N. G. Freier, E. N. Feldman, S. Carrère, and A. Stolyar. "A Plasma Display Window?—The Shifting Baseline Problem in a Technologically Mediated Natural World." *Journal of Environmental Psychology* 28 (2008): 192–199.

Kaplan, R., and S. Kaplan. *The Experience of Nature: A Psychological Perspective*. Cambridge, UK: Cambridge University Press, 1989.

Kaplan, R., and S. Kaplan. "Preference, Restoration, and Meaningful Action in the Context of Nearby Nature." In *Urban Place: Reconnecting with the Natural World*, ed. P. F. Barlett, 271–298. Cambridge, MA: MIT Press, 2005.

Kaplan, S., and R. Kaplan. "Health, Supportive Environments, and the Reasonable Person Model." *American Journal of Public Health* 93, no. 9 (2001): 1484–1489.

Keizer, K., S. Lindenberg, and L. Steg. "The Spreading of Disorder." *Science* 322 (2008): 1681–1684.

Kellert, S. R. *Birthright: People and Nature in the Modern World*. New Haven: Yale University Press, 2012.

Kellert, S. R. *Kinship to Mastery: Biophilia in Human Evolution and Development*. Washington, DC: Island Press, 1997.

Kellert, S. R., J. Heerwagen, and M. Mador, eds. *Biophililc Design: The Theory, Science, and Practice of Bringing Buildings to Life*. Hoboken, NJ: John Wiley & Sons, 2008.

Kellert, S. R., and E. O. Wilson, eds. *The Biophilia Hypothesis*. Washington, DC: Island Press, 1993.

Kelling, G. L., and J. Q. Wilson. "Broken Windows." *The Atlantic*, March 1982, 29–38.

Kennedy, A. M., and M. E. Krasny. "Garden Mosaics: Connecting Science to Community." *Science Teacher* 72, no. 3 (2005): 44–48.

Kindscher, K. "Gardening with Southeast Asian Refugees." 2009. http://kindscher .faculty.ku.edu/wp-content/uploads/2010/10/Kindscher-2009-Gardening-with -SE-Asian-Refugees-2.pdf, accessed July 3, 2014.

King, M. "60 Bronx Youths Are on Right Path." *Daily News*, August 18, 1983.

Kitcher, P. *Vaulting Ambition: Sociobiology and the Quest for Human Nature*. Cambridge, MA: MIT Press, 1987.

Kobori, H. "Current Trends in Conservation Education in Japan." *Biological Conservation* 142 (2009): 1950–1957.

Kobori, H., and R. B. Primack. "Participatory Conservation Approaches for *Satoyama*, the Traditional Forest and Agricultural Landscape of Japan." *Ambio* 32, no. 4 (2003): 307–311.

Krasny, M., and J. Delia. "Natural Area Stewardship as Part of Campus Sustainability." *Journal of Cleaner Production*, May 9, 2014. http://www.sciencedirect .com/science/article/pii/S0959652614003667, accessed July 7, 2014.

Krasny, M., L. Kalbacker, R. C. Stedman, and A. Russ. "Measuring Social Capital among Youth: Applications in Environmental Education." *Environmental Education Research*, 2013, http://www.tandfonline.com/doi/pdf/10.1080/13504622.20 13.843647, accessed July 5, 2014.

Krasny, M., and W.-M. Roth. "Environmental Education for Social-Ecological System Resilience: A Perspective from Activity Theory." *Environmental Education Research* 16, nos. 5–6 (2010): 545–558.

Krasny, M. E., S. R. Crestol, K. G. Tidball, and R. C. Stedman. "New York City's Oyster Gardeners: Memories and Meanings as Motivations of Volunteer Environmental Stewards." *Landscape and Urban Planning* 132 (December 2014): 16–25.

Krasny, M. E., C. Lundholm, E. Lee, S. Shava, and H. Kobori. "Urban Landscapes as Learning Arenas for Sustainable Management of Biodiversity and Ecosystem Services." In *Urbanization, Biodiversity and Ecosystem Services: Challenges and Opportunities*, ed. T. Elmqvist, M. Fragkias, J. Goodness, B. Güneralp, P. J. Marcotullio, R. I. McDonald, S. Parnell, et al., 629–664. New York: Springer, 2013.

Krasny, M. E., K. H. Pace, K. G. Tidball, and K. Helphand. "Greening and Resilience in Military Communities." In *Greening in the Red Zone*, ed. K. G. Tidball and M. E. Krasny, 163–180. New York: Springer, 2014.

Krasny, M. E., A. Russ, K. G. Tidball, and T. Elmqvist. "Civic Ecology Practices: Participatory Approaches to Generating and Measuring Ecosystem Services in Cities." *Ecosystems Services* 7 (March 2014): 177–186.

Krasny, M. E., and K. G. Tidball. "Civic Ecology: A Pathway for Earth Stewardship in Cities." *Frontiers in Ecology and the Environment* 10, no. 5 (2012): 267–273.

Krasny, M. E., and K. G. Tidball. "Community Gardens as Contexts for Science, Stewardship, and Civic Action Learning." *Cities and the Environment* 2, no. 1 (2009). http://digitalcommons.lmu.edu/cgi/viewcontent.cgi?article=1037& context=cate, accessed July 3, 2014.

Krasny, M. E., K. G. Tidball, R. Doyle, and N. Najarian. Garden Mosaics website. N.d. http://communitygardennews.org/gardenmosaics/, accessed July 7, 2014.

Kroth, G. "Guatemala's Garbage Dump Education System." CounterPunch, August 5, 2009. http://www.counterpunch.org/2009/08/05/guatemala-s-garbage -dump-education-system/, accessed July 3, 2014.

Kudryavtsev, A. "Urban Environmental Education and Sense of Place." Ph.D. diss. Ithaca, NY: Cornell University, 2012. http://dx.doi.org/1813/34149, accessed August 1, 2014.

Kudryavtsev, A., M. E. Krasny, and R. C. Stedman. "The Impact of Environmental Education on Sense of Place among Urban Youth." *Ecosphere* 3, no. 4 (2012), http://www.esajournals.org/doi/abs/10.1890/ES11-00318.1, accessed July 7, 2014.

Kudryavtsev, A., R. C. Stedman, and M. E. Krasny. "Sense of Place in Environmental Education." *Environmental Education Research* 18, no. 2 (2011): 229–250.

Kumar, M., and P. Kumar. "Valuation of the Ecosystem Services: A Psycho-Cultural Perspective." *Ecological Economics* 64, no. 4 (2008): 808–819.

Kuo, F. E., M. Bacaicoa, and W. C. Sullivan. "Transforming Inner-City Landscapes: Trees, Sense of Safety, and Preference." *Environment and Behavior* 30 (1998): 28–59.

Kuo, F. E., and W. C. Sullivan. "Environment and Crime in the Inner City: Does Vegetation Reduce Crime?" *Environment and Behavior* 33 (2001): 343–367.

Kuo, F. E., W. C. Sullivan, R. L. Coley, and L. Brunson. "Fertile Ground for Community: Inner-City Neighborhood Common Spaces." *American Journal of Community Psychology* 26, no. 6 (1998): 823–855.

Kurlansky, M. *The Big Oyster: History on the Half Shell.* New York: Ballantine Books, 2006.

Lave, J., and E. Wenger. *Situated Learning.* Cambridge, UK: Cambridge University Press, 1991.

Lawson, L. J. *City Bountiful: A Century of Community Gardening in America.* Berkeley: University of California Press, 2005.

Leca, B., J. Battilana, and E. Boxenbaum. "Agency and Institutions: A Review of Institutional Entrepreneurship." Working paper 08-096, Harvard Business School, 2008. http://egateg.usaidallnet.gov/sites/default/files/Review%20of%20Institutional%20Entrepreneurship.pdf, accessed July 7, 2014.

Lee, E. "Reconstructing Village Groves after a Typhoon in Korea." In *Greening in the Red Zone*, ed. K. G. Tidball and M. E. Krasny, 159–162. New York: Springer, 2014.

Lee, E., and M. E. Krasny. "The Role of Social Learning for Social-Ecological Systems in Korean Village Groves Restoration." Forthcoming in *Ecology and Society.*

Lee, M. J. "Possibilities for the Emergence of Civic Ecology Practices in Response to Social-Ecological Disturbance: The Case of Nuisance Chironomids in Singapore." Ithaca, NY: Civic Ecology Lab, Cornell University, 2013. http://civeco.files .wordpress.com/2013/09/2013-lee.pdf, accessed July 7, 2014.

Leibowitz, L. "Humans and Other Animals: A Perspective on Perspectives." *Dialectical Anthropology* 9, nos. 1–4 (1985): 127–142.

Lemann, N. *The Promised Land: The Great Black Migration and How It Changed America.* New York: Random House, 1991.

Leopold, A. *A Sand County Almanac.* New York: Oxford University Press, 1949.

Levin, S. "Self-Organization and the Emergence of Complexity in Ecological Systems." *BioScience* 55, no. 12 (2005): 1075–1079.

Levin, S., and W. Clark, eds. "Toward a Science of Sustainability: Report from Toward a Science of Sustainability Conference." CID Working Paper No. 196. Cambridge, MA: Center for International Development, Harvard University, May 2010. http://www.hks.harvard.edu/var/ezp_site/storage/fckeditor/file/pdfs/ centers-programs/centers/cid/publications/faculty/wp/196.pdf, accessed August 2, 2014.

Lewis, C. "Public Housing Gardens: Landscapes for the Soul." In *Landscape for Living*, 277–282. Washington, DC: USDA Yearbook of Agriculture, 1972. http:// naldc.nal.usda.gov/download/IND79000696/PDF, accessed July 7, 2104.

Lewis, C., and S. Sturgill. "Comment: Healing in the Urban Environment: A Person/Plant Viewpoint." *Journal of the American Planning Association* 45, no. 3 (1979): 330–338.

Lewontin, R. C., S. P. R. Rose, and L. J. Kamin. *Not in Our Genes: Biology, Ideology, and Human Nature.* New York: Pantheon, 1984.

Light, A. "Urban Ecological Citizenship." *Journal of Social Philosophy* 34, no. 1 (2003): 44–63.

Lindow, M. "Mountain of Hope." *Sierra Magazine*, March/April 2005. http:// vault.sierraclub.org/sierra/200503/profile.asp, accessed July 7, 2014.

Liu, J., T. Dietz, S. R. S. R. Carpenter, M. Alberti, C. Folke, E. Moran, A. N. Pell, et al. "Complexity of Coupled Human and Natural Systems." *Science* 317 (2007): 1513–1516.

Livingston, J. A. *The Fallacy of Wildlife Conservation.* Toronto: McClelland and Stewart, 1981.

Lochner, K., I. Kawachi, and B. P. Kennedy. "Social Capital: A Guide to Its Measurement." *Health & Place* 5 (1999): 259–270.

Löf, A. "Exploring Adaptability through Learning Layers and Learning Loops." *Environmental Education Research* (special issue on Resilience in Social-Ecological Systems: The Roles of Learning and Education; ed. M. E. Krasny, C. Lundholm, and R. Plummer) 15, nos. 5–6 (2010): 529–543.

Long, J. "Detroit's Demise Sets Up Rebirth from Grassroots." *Bloomberg Businessweek*, April 6, 2011. http://www.bloomberg.com/news/2011-04-06/detroit-s -fall-sets-up-rebirth-from-grassroots-commentary-by-joshua-long.html, accessed August 1, 2014.

Longhurst, J. "Mennonites, Sikhs Work Together on Garden Project." May 26, 2012. http://www.winnipegfreepress.com/arts-and-life/life/faith/mennonites-sikhs-work -together-on-garden-project-planting-seeds-of-harmony-154386235.html, accessed July 7, 2014.

Louv, R. *Last Child in the Woods: Saving Our Children from Nature-Deficit Disorder.* New York: Algonquin Books, 2006.

Lovett, K. "Gov. Cuomo: Sandy as Bad as Anything I've Experienced in New York." *Daily News*, October 30, 2012. http://www.nydailynews.com/blogs/dailypolitics/gov-cuomo-sandy-bad-experienced-new-york-blog-entry-1.1692632, accessed July 7, 2014.

Luthar, S. S., D. Cicchetti, and B. Becker. "The Contruct of Resilience: A Critical Evaluation and Guidelines for Future Work." *Child Development* 71, no. 3 (2000): 543–562.

Lyson, T. *Civic Agriculture: Reconnecting Farm, Food, and Community.* Lebanon, NH: Tufts University Press, 2004.

Madrigal, A. C. "The Sacrificial Landscape of True Detective." *The Atlantic*, March 7, 2014. http://www.theatlantic.com/technology/archive/2014/03/the -sacrifical-landscape-of-em-true-detective-em/284302/, accessed July 7, 2014.

Maguire, S., C. Hardy, and T. B. Lawrence. "Institutional Entrepreneurship in Emerging Fields: HIV/AIDS Treatment Advocacy in Canada." *Academy of Management Journal* 47, no. 5 (2004): 657–679.

Malakoff, D. "Research, Ecological Restoration Get Unprecedented Windfall from BP Oil Spill Criminal Settlement." *Science (AAAS)*, November 16, 2012. http://news.sciencemag.org/people-events/2012/11/research-ecological-restoration -get-unprecedented-windfall-bp-oil-spill, accessed July 6, 2014.

Mancabelli, A. "Symbolic Meaning and Emotion in Japanese Ethnozoology." Paper presented at the University of Alaska Anchorage Annual Ethnobiological Conference, May 2002. https://www.yumpu.com/en/document/view/6907189 /symbolic-meaning-and-emotion-in-japanese-ethnozoology, accessed July 3, 2014.

Manzo, L. C., and D. D. Perkins. "Finding Common Ground: The Importance of Place Attachment to Community Participation and Planning." *Journal of Planning Literature* 20, no. 4 (2006): 335–350.

Marshall, K., J. Hamlin, M. Armstrong, J. Mendoza, C. Lee, D. Pieri, R. Rivera, et al. "Science for a Social Revolution: Ecologists Entering the Realm of Action." *Bulletin of the Ecological Society of America* 92, no. 3 (2011): 241–243.

Martin, J. *Genius of Place: The Life of Frederick Law Olmsted*. Cambridge, MA: Da Capo Press, 2011.

Masten, A. S., and J. Obradovic. "Disaster Preparation and Recovery: Lessons from Research on Resilience in Human Development." *Ecology and Society* 13, no. 1 (2008), http://www.ecologyandsociety.org/vol13/iss1/art9/, accessed July 7, 2014.

McCay, B. J., and A. C. Finlayson. "The Political Ecology of Crisis and Institutional Change: The Case of the Northern Cod." Paper presented at the Annual Meeting of the American Anthropological Association, Washington, DC, 1995.

McIntosh, R. J., J. A. Tainter, and S. K. McIntosh. "Climate, History, and Human Action." In *The Way the Wind Blows: Climate, History, and Human Action*, ed. R. J. McIntosh, J. A. Tainter, and S. K. McIntosh, 1–42. New York: Columbia University Press, 2000.

McKnight, J. L., and J. P. Kretzman. "Mapping Community Capacity." Evanston, IL: Institute for Policy Research, 1996.

MEA. *Millennium Ecosystem Assessment: Ecosystems and Human Well-Being: Synthesis*. Washington, DC: Island Press, 2005.

Meadows, D., and P. Marshall. "Dancing with Systems." *Whole Earth* 106 (2001): 58–63.

Medlar, P., and H. Sklar. *Streets of Hope: The Fall and Rise of an Urban Neighborhood*. Boston: South End Press, 1994.

Meijerink, S., and D. Huitema. "Policy Entrepreneurs and Change Strategies: Lessons from Sixteen Case Studies of Water Transitions around the Globe." *Ecology and Society* 15, no. 2 (2010), http://www.ecologyandsociety.org/vol15/iss2/art21/, accessed July 7, 2014.

Merino, L. "Andre's River." 2012. http://vimeo.com/49071323, accessed July 7, 2014.

Miles, I., W. Sullivan, and F. Kuo. "Ecological Restoration Volunteers: The Benefits of Participation." *Urban Ecosystems* 2 (1998): 27–41.

Mitchell, J. *The Bottom of the Harbor.* New York: Random House, Inc., 1951.

Monarch Watch. N.d. http://www.monarchwatch.org/, accessed July 3, 2014.

Monti, D. "Experimental Reefs Could Add Life to Bay: No Fluke." December 11, 2013. http://www.warwickonline.com/stories/Experimental-reefs-could-add-life-to-Bay,88203?print=1, accessed July 7, 2014.

Moore, E. "Growing Hope: How Urban Gardens Are Empowering War-Affected Liberians and Harvesting a New Generation of City Farmers." In *Greening in the Red Zone,* ed. K. G. Tidball and M. E. Krasny, 411–417. New York: Springer, 2014.

Moore, M.-L., and F. Westley. "Surmountable Chasms: Networks and Social Innovation for Resilient Systems." *Ecology and Society* 16, no. 1 (2011), http://www.ecologyandsociety.org/vol16/iss1/art5/, accessed July 6, 2014.

Moser, E. "Seeding the City." N.d. http://www.evemosher.com/2011/seedingthecity/, accessed July 7, 2014.

Moskovits, D. K., C. Fialkowski, G. M. Mueller, T. A. Sullivan, J. Rogner, and E. McCance. "Chicago Wilderness: A New Force in Urban Conservation." *Annals of the New York Academy of Sciences* 1023 (2004): 215–236.

Mulgan, G. "Social Innovation: What It Is, Why It Matters and How It Can Be Accelerated." London: Young Foundation, 2006. http://dsi.britishcouncil.org.cn/images/SocialInnovation.pdf, accessed July 7, 2014.

Mulvaney, K. "Time to Bring Back New York's Oysters." November 14, 2012. http://news.discovery.com/earth/new-york-needs-new-oysters-121114.htm, accessed July 7, 2014.

Muro, M., and P. Jeffrey. "A Critical Review of the Theory and Application of Social Learning in Participatory Natural Resource Management Processes." *Journal of Environmental Planning and Management* 51, no. 3 (2008): 325–344.

Næss, A. "The Deep Ecology 'Eight Points' Revisited." In *The Selected Works of Arne Næss,* ed. Harold Glasser and Alan Drengson, 57–66. Dordrecht, The Netherlands: Springer, 2005.

Næss, A. "Self-Realization: An Ecological Approach to Being in the World." In *The Selected Works of Arne Næss,* ed. Harold Glasser and Alan Drengson, 515–530. Dordrecht, The Netherlands: Springer, 2005.

Nathe, M. "Guatemala City's Dirtiest Secret." *Endeavors,* January 14, 2008. http://endeavors.unc.edu/win2008/guatemala.php, accessed August 1, 2014.

National Audubon Society. "Toyota Together Green: Ben Haberthur." N.d. http://www.togethergreen.org/fellows/fellow/benjamin-haberthur, accessed July 7, 2014.

National Recreation and Park Foundation. "Marvin Gaye Park Backgrounder." N.d. http://www.nrpa.org/uploadedFiles/Events/MGP%20Backgrounder.pdf?n=1508, accessed July 3, 2014.

National Wildlife Federation. "Earth Conservation Corps (ECC): Reclaiming Two of America's Most Endangered Resources—Our Youth and Our Environment." N.d. http://www.nwf.org/news-and-magazines/media-center/faces-of-nwf/ecc.aspx, accessed July 3, 2014.

Nelson Mandela Foundation. *A Prisoner in the Garden*. New York: Viking Studio, Penguin Group, 2006.

Nesper, C. "Legitimate Protection or Tactful Abandonment: Can Recent California Legislation Sustain the San Francisco Bay Area's Public Lands?" *Golden Gate University Environmental Law Journal* 6, no. 1 (2012): 153–181.

Nisbet, E. K., and J. M. Zelenski. "Underestimating Nearby Nature: Affective Forecasting Errors Obscure the Happy Path to Sustainability." *Psychological Science* 22, no. 9 (2011): 1101–1106.

NOAA. "Military Veterans Help Rebuild Fisheries." May 22, 2013. http://www.habitat.noaa.gov/highlights/vetsrestorationvideo.html, accessed July 4, 2014.

Noddings, N. "On Community." *Educational Theory* 46, no. 3 (1996): 245–267.

NY/NJ Baykeeper. N.d. http://www.nynjbaykeeper.org/, accessed July 3, 2014.

NYC Department of Parks & Recreation. "The Community Garden Movement: Green Guerrillas Gain Ground." N.d. http://www.nycgovparks.org/about/history/community-gardens/movement, accessed July 7, 2014.

NYC Parks. "Concrete Plant Park." N.d. http://www.nycgovparks.org/park-features/concrete-plant-park/planyc, accessed July 7, 2014.

NYC Service. "NYC Service." http://www.nycservice.org/.

O'Brien, O. "Cape Flats, Cape Town." May 24, 2012. http://blogs.newschool.edu/epsm/2012/05/24/cape-flats-cape-town/, accessed July 7, 2014.

Okvat, H., and A. Zautra. "Community Gardening: A Parsimonious Path to Individual, Community, and Environmental Resilience." *American Journal of Community Psychology* 47 (2011): 374–387.

Okvat, H., and A. Zautra. "Sowing Seeds of Resilience: Community Gardening in a Post-Disaster Context." In *Greening in the Red Zone*, ed. K. G. Tidball and M. E. Krasny. New York: Springer, 2014: 73–90.

Olmsted, F. L., and C. Vaux. "A Review of Recent Changes, and Changes which have been Projected, in the Plans of the Central Park, by the Landscape Architects, 1872." In *F. L. Olmsted, Landscape Architect, 1822–1903*, vol. 2: *Central Park as a Work of Art and as a Great Municipal Enterprise 1853–1895*, 3rd ed., ed. Frederick Law Olmsted Jr. and Theodora Kimball (Bronx, NY: Benjamin Blom, Inc., 1970).

Olsson, P., C. Folke, and F. Berkes. "Adaptive Co-Management for Building Resilience in Social-Ecological Systems." *Environmental Management* 34, no. 1 (2004): 75–90.

Olsson, P., C. Folke, V. Galaz, T. Hahn, and L. Schultz. "Enhancing the Fit through Adaptive Comanagement: Creating and Maintaining Bridging Functions for Matching Scales in the Kristianstads Vattenrike Biosphere Reserve Sweden." *Ecology and Society* 12, no. 1 (2007), http://www.ecologyandsociety.org/vol12/iss1/art28/, accessed September 17, 2010.

Ostrom, E. *Governing the Commons.* New York: Cambridge University Press, 1990.

Ostrom, E. "Polycentric Systems for Coping with Collective Action and Global Environmental Change." *Global Environmental Change* 20 (2010): 550–557.

Oxford Dictionary. "Bricolage." N.d. http://www.oxforddictionaries.com/definition/english/bricolage, accessed July 7, 2014.

Page, N., and C. E. Czuba. "Empowerment: What Is It?" *Journal of Extension* 37, no. 5 (1999). http://www.joe.org/joe/1999october/comm1.php, accessed July 4, 2014.

Pahl-Wostl, C. "The Importance of Social Learning in Restoring the Multifunctionality of Rivers and Floodplains." *Ecology and Society* 11, no. 1 (2006), http://www.ecologyandsociety.org/vol11/iss1/art10/, accessed July 5, 2014.

Pahl-Wostl, C., M. Craps, A. Dewulf, E. Mostert, D. Tabara, and T. Tailleu. "Social Learning and Water Resources Management." *Ecology and Society* 12, no. 2 (2007). http://www.ecologyandsociety.org/vol12/iss2/art5, accessed July 7, 2014.

Palamar, C. R. "The Justice of Ecological Restoration: Environmental History, Health, Ecology, and Justice in the United States." *Human Ecology Review* 15, no. 1 (2008): 82–94.

Palmer, M. A. "Socioenvironmental Sustainability and Actionable Science." *BioScience* 62, no. 1 (2012): 5–6.

Panetta, R. G. *Westchester: The American Suburb.* New York: Fordham University Press, 2006.

Park, B. J., Y. Tsunetsugu, T. Kasetani, T. Kagawa, and Y. Miyazaki. "The Physiological Effects of Shinrin-Yoku (Taking in the Forest Atmosphere or Forest Bathing): Evidence from Field Experiments in 24 Forests across Japan." *Environmental Health and Preventive Medicine* 15 (2010): 18–26.

Pataki, D. E., M. M. Carreiro, J. Cherrier, N. E. Grulke, E. Jennings, S. Pincet, R. V. Pouyat, T. H. Whitlow, and W. C. Zipperer. "Coupling Biogeochemical Cycles in Urban Environments: Ecosystem Services, Green Solutions, and Misconceptions." *Frontiers in Ecology and the Environment* 9, no. 1 (2011): 27–36.

Pelling, M., and K. Dill. "Disaster Politics: Tipping Points for Change in the Adaptation of Sociopolitical Regimes." *Progress in Human Geography* 34, no. 1 (2010): 21–37.

Pelling, M., and D. Manuel-Navarrete. "From Resilience to Transformation: The Adaptive Cycle in Two Mexican Urban Centers." *Ecology and Society* 16, no. 2 (2011), http://www.ecologyandsociety.org/vol16/iss2/art11/, accessed August 2, 2014.

Peterson, M. J., D. M. Hall, A. M. Feldpausch-Parker, and T. R. Peterson. "Obscuring Ecosystem Function with Application of the Ecosystem Services Concept." *Conservation Biology* 24, no. 1 (2009): 113–119.

Pitt, B., and T. Boulle. "Together: Thinking and Practice of Urban Nature Conservators." Cape Town: SANBI Cape Flats Nature, 2010. http://reworrr.blogspot.com/2013/11/combat-wounded-veterans-replant-coral.html, accessed July 7, 2014.

PlaNYC. "MillionTreesNYC." NYC Parks and NY Restoration Project. N.d. http://www.milliontreesnyc.org/html/home/home.shtml, accessed July 7, 2014.

Pollock, M. "Cut Casino Tax Rate to Encourage Investment." *The Press of Atlantic City.* 2014. http://www.pressofatlanticcity.com/opinion/commentary/michael-pollock-cut-casino-tax-rate-to-encourage-investment/article_8373532b-d200-5877-8b60-ee364b1ed920.html?mode=jqm, accessed July 2, 2014.

Poole, K. "Civitas Oecologie: Infrastructure in the Ecological City." In *The Harvard Architect Review*, ed. T. Genovese, L. Eastley, and D. Snyder, 126–142. New York: Princeton Architectural Press, 1998.

Portes, A. "Social Capital: Its Origins and Applications in Modern Sociology." *Annual Review of Sociology* 24 (1998): 1–24.

Portes, A., and P. Landolt. "Social Capital: Promise and Pitfalls of Its Role in Development." *Journal of Latin American Studies* 32, no. 2 (2000): 529–547.

Powell, M. "Bronx River Sankofa: Eco-Cultural History in New York City!" April 8, 2004. http://bronxriversankofa.wordpress.com/, accessed July 7, 2014.

Powell, M. "West Farms Rapids Park: A History," February 25, 2014. http://bronxriversankofa.wordpress.com/2014/02/25/west-farms-rapids-park-a-history/, accessed July 7, 2014.

Primack, R. B., H. Kobori, and S. Mori. "Dragonfly Pond Restoration Promotes Conservation Awareness in Japan." *Conservation Biology* 14, no. 5 (2000): 1153–1154.

Putnam, R. B. "Bowling Alone: America's Declining Social Capital." *Journal of Democracy* 6, no. 1 (1995): 65–78.

Putnam, R. B. *Bowling Alone: The Collapse and Revival of American Community.* New York: Simon & Schuster, 2000.

Rappaport, J. "In Praise of Pardox: A Social Policy of Empowerment over Prevention." *American Journal of Community Psychology* 9, no. 1 (1981): 1–25.

Reed, C. "The Ecological (and Urbanistic) Agency of Infrastructure." In *Deconstruction/Reconstruction: The Cheonggyecheon Restoration Project in Seoul*, ed. J. Busquets, 35–46. Cambridge, MA: Harvard University Gradauate School of Design, 2011.

Reef Ball Foundation. "The Reef Ball Foundation-Designed Artificial Reefs." N.d. http://www.reefball.org/, accessed July 7, 2014.

Relf, P. D. "The Therapeutic Values of Plants." *Pediatric Rehabilitation* 8, no. 3 (2005): 235–237.

Revkin, A. "Students Press the Case for Oysters as New York's Surge Protector." *New York Times*, November 12, 2012. http://dotearth.blogs.nytimes .com/2012/11/12/students-press-the-case-for-oysters-as-new-yorks-surge -protector/, accessed July 7, 2014.

Reyers, B., R. Biggs, G. S. Cumming, T. Elmqvist, A. P. Hejnowicz, and S. Polasky. "Getting the Measure of Ecosystem Services: A Social-Ecological Approach." *Frontiers in Ecology and the Environment* 11, no. 5 (2013): 268–273.

Richardson, G. P. "Problems with Causal Loop Diagrams." *System Dynamics Review* 2, no. 2 (1986). http://www.systems-thinking.org/intst/d-3312.pdf, accessed July 7, 2014.

Richardson, G. P. "Systems Dynamics in the Elevator." N.d. http://www.stewardshipmodeling.com/System%2520Dynamics%2520in%2520the%2520Elevator .htm/20Elevator.htm, accessed July 7, 2014.

Riker, J. V., and K. E. Nelson. "What Works to Strengthen Civic Engagement in America: A Guide to Local Action and Civic Innovation. College Park, MD: The Democracy Collaborative-Knight Foundation Civic Engagement Project, 2003. http://www.scribd.com/doc/13332876/Knight-Synthesis-Report-Final, accessed July 7, 2014.

Roberston, D. P., and R. B. Hull. "Public Ecology: An Environmental Science and Policy for Global Society." *Environmental Science & Policy* 6 (2003): 399–410.

Rocking the Boat (RTB). N.d. http://www.rockingtheboat.org/, accessed August 2, 2014.

Rockström, J., W. Steffen, K. Noone, Å. Persson, I. F. S. Chapin, E. Lambin, T. M. Lenton, et al. "Planetary Boundaries: Exploring the Safe Operating Space for Humanity." *Ecology and Society* 14, no. 2 (2009), http://www.ecologyandsociety .org/vol14/iss2/art32, accessed July 3, 2014.

Rogers, D. "Planners Push to Tear out Elevated I-10 over Claiborne." *Times-Picayune*, July 11, 2009. http://floor8.info/nola/20090700/090711_proposal_to _tear_down_i_10_over_claiborne.html, accessed July 7, 2014.

Rolston, H. "Nature for Real: Is Nature a Social Construction?" In *The Philosophy of the Environment*, ed. T. D. J. Chappell, 38–64. Edinburgh: University of Edinburgh Press, 1997.

Ruitenbeek, J., and C. Cartier. "The Invisible Wand: Adaptive Co-management as an Emergent Strategy in Complex Bio-economic Systems." Occasional paper. Bogor, Indonesia: Center for International Forestry Research, October 2001. http://www.cifor.org/publications/pdf_files/OccPapers/OP-034.pdf, accessed July 7, 2014.

Sachs, J. D. *The End of Poverty: Economic Possibilities for Our Time*. New York: Penguin Press, 2005.

Saguaro Seminar. "Social Capital Community Benchmark Survey: Executive Summary." Cambridge, MA: Harvard University, n.d. http://www.hks.harvard.edu/ saguaro/communitysurvey/docs/exec_summ.pdf, accessed July 3, 2014.

Saldivar, L., and M. E. Krasny. "The Role of NYC Latino Community Gardens in Community Development, Open Space, and Civic Agriculture." *Agriculture and Human Values* 21 (2004): 399–412.

Sampson, R. J., D. McAdam, H. MacIndoe, and S. Weffer-Elizondo. "Civil Society Reconsidered: The Durable Nature and Community Structure of Collective Civic Action." *American Journal of Sociology* 111, no. 3 (2005): 673–714.

Sampson, R. J., S. W. Raudenbush, and F. Earls. "Neighborhoods and Violent Crime: A Multilevel Study of Collective Efficacy." *Science* 277, no. 5328 (1997): 918–924.

Sassen, S., and N. Dotan. "Delegating, Not Returning, to the Biosphere: How to Use the Multi-Scalar and Ecological Properties of Cities." *Global Environmental Change* 21 (2011): 823–834.

Saw Mill River Coalition. "Daylighting the Saw Mill River in Yonkers." N.d. http://www.sawmillrivercoalition.org/whats-happening/daylighting-the-saw-mill -river-in-yonkers/, accessed July 7, 2014.

Schaefer, V. "Alien Invasions, Ecological Restoration in Cities and the Loss of Ecological Memory." *Restoration Ecology* 17, no. 2 (2009): 171–176.

Scheffer, M., and F. Westley. "The Evolutionary Basis of Rigidity: Locks in Cells, Minds, and Society." *Ecology and Society* 12, no. 2 (2007), http://www.ecologyan-dsociety.org/vol12/iss2/art36/, accessed July 6, 2014.

Schilling, J., and J. Logan. "Greening the Rust Belt: A Green Infrastructure Model for Right Sizing America's Shrinking Cities." *Journal of the American Planning Association* 74, no. 4 (2008): 451–466.

Schlesinger, W. H. "Translational Ecology." *Science* 329 (2010): 609.

Schlosberg, D., and J. S. Dryzek. "Political Strategies of American Environmentalism: Inclusion and Beyond." *Society & Natural Resources* 15 (2002): 787–804.

Schusler, T. M., D. J. Decker, and M. J. Pfeffer. "Social Learning for Collaborative Natural Resource Management." *Society & Natural Resources* 15 (2003): 309–326.

Schwartz, M. "On Decimated Shore, a Second Life for Christmas Trees." *New York Times*, February 3, 2013, http://www.nytimes.com/2013/02/04/nyregion/ on-decimated-shore-a-second-life-for-christmas-trees.html?_r=0, accessed July 10, 2014.

Seeley, T. D. *Honeybee Democracy*. Princeton: Princeton University Press, 2010.

Segerstråle, U. *Defenders of the Truth: The Battle for Science in the Sociobiology Debate and Beyond*. Oxford: Oxford University Press, 2000.

Seyfang, G., and A. Haxeltine. "Growing Grassroots Innovations: Exploring the Role of Community-Based Initiatives in Governing Sustainable Energy Transitions." *Environment and Planning. C, Government & Policy* 30 (2012): 381–400.

Shava, S., and M. Mentoor. "Turning Degraded Open Space into a Community Asset: The Soweto Mountain of Hope Greening Case Story." In *Greening in the Red Zone*, ed. K. G. Tidball and M. E. Krasny, 91–94. New York: Springer, 2014.

Sheffield, H. "How One Unemployed Bronxite Transformed Her Block." October 20, 2011. http://hazelsheffield.com/tag/nilka-martell/, accessed February 26, 2014.

Sheffield, H. "Turning Unemployment into Time Well Spent." *The Bronx Ink*, November 3, 2011. http://bronxink.org/2011/11/03/19123-turning-unemployment-into-time-well-spent/, accessed July 7, 2014.

Shirk, J., H. Ballard, C. Wilderman, T. Phillips, A. Wiggins, R. Jordan, E. McCallie, M. Minarchek, B. Lewenstein, M. E. Krasny, and R. Bonney. "Public Participation in Scientific Research: A Framework for Deliberate Design." *Ecology and Society* 17, no. 2 (2012). http://www.ecologyandsociety.org/vol17/iss2/art29/, accessed July 7, 2014.

Shutkin, W. A. *The Land That Could Be: Environmentalism and Democracy in the Twenty-First Century*. Cambridge, MA: MIT Press, 2001.

Sideris, L. H. *Environmental Ethics, Ecological Theology, and Natural Selection*. New York: Columbia University Press, 2003.

Silva, P., and M. E. Krasny. "Outcomes Monitoring within Civic Ecology Practices in the NYC Region." Ithaca, NY: Civic Ecology Lab, Cornell University, 2013. http://civeco.files.wordpress.com/2013/09/silva-2013.pdf, accessed July 7, 2014.

Silva, P., and M. E. Krasny. "Parsing Participation: Models of Engagement for Outcomes Monitoring in Urban Stewardship." *Local Environment: The International Journal of Justice and Sustainability*, 2014. http://dx.doi.org/10.1080/135 49839.2014.929094.

Sirianni, C. *Investing in Democracy: Engaging Citizens in Collaborative Governance*. Washington, DC: Brookings Institution Press, 2009.

Sirianni, C., and L. A. Friedland. *Civic Innovation in America: Community Empowerment, Public Policy, and the Movement for Civic Renewal*. Berkeley: University of California Press, 2001.

Sirianni, C., and L. A. Friedland. *The Civic Renewal Movement: Community Building and Democracy in the United States*. Dayton, OH: Charles F. Kettering Foundation, 2005.

Slater, R. J. "Urban Agriculture, Gender and Empowerment: An Alternative View." *Development Southern Africa* 18, no. 5 (2001): 635–650.

Slater, T. "The Resilience of Neoliberal Urbanism." January 28, 2014. http://www.opendemocracy.net/opensecurity/tom-slater/resilience-of-neoliberal-urbanism, accessed July 7, 2014.

Sledge, M. "Hurricane Sandy Shows We Need to Prepare for Climate Change, Cuomo and Bloomberg Say." *Huffington Post*, October 30, 2012. http://www.huffingtonpost.com/2012/10/30/hurricane-sandy-cuomo-bloomberg-climate-change_n_2043982.html, accessed July 7, 2014.

Smith, T. W. "Civic Ecology: A Community Systems Approach to Sustainability." *Oregon Planners' Journal*, March/April 2008, 10–12, 14–15.

Stedman, R. C. "Toward a Social Psychology of Place: Predicting Behavior from Place-Based Cognitions, Attitude, and Identity." *Environment and Behavior* 34 (2002): 561–581.

Stedman, R. C., and M. Ingalls. "Topophilia, Biophilia and Greening in the Red Zone." In *Greening in the Red Zone*, ed. K. G. Tidball and M. E. Krasny, 129–144. New York: Springer, 2014.

Steinzor, R. I. "The Corruption of Civic Environmentalism." *The Environmental Law Reporter* 30, no. ELR 10909 (2000). http://elr.info/news-analysis/30/10909/corruption-civic-environmentalism, accessed July 7, 2014.

Stern, P. C. "The Value Basis of Environmental Concern." *Journal of Social Issues* 50, no. 3 (1994): 65–84.

Stevens, W. K. *Miracle under the Oaks: The Revival of Nature in America*. New York: Pocket Books, 1995.

Stoker, G. "Governance as Theory: Five Propositions." *International Social Science Journal* 50, no. 155 (1998): 17–28.

Strauss, E. "Urban Pollinators and Community Gardens." *Cities and the Environment* 2, no. 1 (2009). http://digitalcommons.lmu.edu/cate/vol2/iss1/1/, accessed July 3, 2014.

Stuart, S. M. "Lifting Spirits: Creating Gardens in Domestic Violence Shelters." In *Urban Place: Reconnecting with the Natural World*, ed. P. F. Barlett, 61–88. Cambridge, MA: MIT Press, 2005.

Sullivan, P. "Getting Its Groove Back." *Washington Post*, April 2, 2006. http://www.washingtonpost.com/wp-dyn/content/article/2006/04/01/AR2006040101105.html, accessed July 3, 2014.

Sullivan, W., F. Kuo, and S. F. DePooter. "The Fruit of Urban Nature: Vital Neighborhood Spaces." *Environment and Behavior* 36, no. 5 (2004): 678–700.

Svendsen, E., and L. Campbell. "Land-Markings: 12 Journeys through 9/11 Living Memorials." Newtown Square, PA: U.S. Department of Agriculture, Forest Service, Northern Research Station, 2006. http://www.livingmemorialsproject.net/DOWNLOADS/NRS-INF-1-06.pdf, accessed July 5, 2014.

Svendsen, E. S., and L. Campbell. "Community-Based Memorials to September 11, 2001: Environmental Stewardship as Memory Work." In *Greening in the Red Zone*, ed. K. G. Tidball and M. E. Krasny, 339–355. New York: Springer, 2014.

Svendsen, E.S., and L. Campbell. "Living Memorials Project: Year 1 Social and Site Assessment." U.S. Forest Service General Technical Report NE-333, 2005.

Svendsen, E. S., and L. Campbell. "Urban Ecological Stewardship: Understanding the Structure, Function and Network of Community-Based Land Management." *Cities and the Environment* 1, no. 1 (2008): 1–31.

Tainter, J. A. "Global Change, History, and Sustainability." In *The Way the Wind Blows: Climate, History, and Human Action*, ed. R. J. McIntosh, J. A. Tainter, and S. K. McIntosh, 331–356. New York: Columbia University Press, 2000.

Taylor, K. "After High Line's Success, Other Cities Look Up." *New York Times*, July 14, 2010. http://www.nytimes.com/2010/07/15/arts/design/15highline .html?_r=1&, accessed July 7, 2014.

TEEB. "TEEB—the Economics of Ecosystems and Biodiversity for Local and Regional Policy Makers." Report for the United National Environment Program. 2010. http://www.teebweb.org/publication/teeb-for-local-and-regional-policy-makers-2, accessed July 3, 2014.

Tidball, K. G. "Combat Wounded Veterans Replant Coral Fragments." November 30, 2013. http://reworrr.blogspot.com/2013/11/combat-wounded-veterans -replant-coral.html, accessed July 7, 2014.

Tidball, K. G. "Greening in the Red Zone: Valuing Community-Based Ecological Restoration in Human Vulnerability Contexts." Ph.D. diss. Ithaca, NY: Cornell University, 2012. http://dspace.library.cornell.edu/handle/1813/31450, accessed July 3, 2014.

Tidball, K. G. "Mechanisms of Resilience & Other 'Re-Words' in Urban Greening." April 24, 2013. http://www.thenatureofcities.com/2013/04/24/mechanisms -of-resilience-other-re-words-in-urban-greening/, accessed February 23, 2014.

Tidball, K.G. "Outdoor Recreation, Restoration and Healing for Returning Combatants." September 23, 2013. http://www.thenatureofcities.com/2013/09/23/outdoor -recreation-restoration-and-healing-for-returning-combatants/, accessed July 3, 2014.

Tidball, K. G. "Peace Research and Greening in the Red Zone: Community-Based Ecological Restoration to Enhance Resilience and Transitions toward Peace." In *Expanding Peace Ecology: Peace, Security, Sustainability, Equity and Gender: Perspectives of IPRA's Ecology and Peace Commission*, ed. U. O. Spring, H. G. Brauch, and K. G. Tidball, 63–83. Berlin: Springer-Verlag, 2014.

Tidball, K. G. "Seeing the Forest for the Trees: Hybridity and Social-Ecological Symbols, Rituals and Resilience in Post-Disaster Contexts." Forthcoming in *Ecology and Society* (special issue on Exploring Social-Ecological Resilience through the Lens of the Social Sciences: Contributions, Critical Reflections and Constructive Debate).

Tidball, K. G. "Trees and Rebirth: Social-Ecological Symbols and Rituals in the Resilience of Post-Katrina New Orleans." In *Greening in the Red Zone*, ed. K. G. Tidball and M. E. Krasny, 257–296. New York: Springer, 2014.

Tidball, K. G. "Urgent Biophilia: Human-Nature Interactions and Biological Attractions in Disaster Resilience." *Ecology and Society* 17, no. 2 (2012). http:// www.ecologyandsociety.org/vol17/iss2/art5/, accessed July 7, 2014.

Tidball, K. G. "Urgent Biophilia: Human-Nature Interactions in Red Zone Recovery and Resilience." In *Greening in the Red Zone*, ed. K. G. Tidball and M. E. Krasny, 53–73. New York: Springer, 2014.

Tidball, K. G., A. Atkipis, and R. C. Stedman. "Greening and Positive Feedback: Cultural, Cognitive and Environmental Factors Influencing Willingness to Invest in Local Ecological Infrastructure through Greening." Forthcoming in *Human Ecology*.

Tidball, K. G., M. Bain, and T. Elmqvist. "Detecting Virtuous Cycles That Confer Resilience in Disrupted Social-Ecological Systems." Forthcoming in *Ecology and Society*.

Tidball, K. G., and M. E. Krasny. "From Risk to Resilience: What Role for Community Greening and Civic Ecology in Cities?" In *Social Learning Towards a More Sustainable World*, ed. A. E. J. Wals, 149–164. Wagengingen, The Netherlands: Wagengingen Academic Press, 2007.

Tidball, K. G., and M. E. Krasny. *Greening in the Red Zone*. New York: Springer, 2014.

Tidball, K. G., and M. E. Krasny. "Toward an Ecology of Environmental Education and Learning." *Ecosphere* 2, no. 2 (2011). http://www.esajournals.org/doi /pdf/10.1890/ES10-00153.1, accessed July 3, 2014.

Tidball, K. G., M. E. Krasny, E. S. Svendsen, L. Campbell, and K. Helphand. "Stewardship, Learning, and Memory in Disaster Resilience." *Environmental Education Research* 16, nos. 5–6 (2010): 591–609.

Tidball, K. G., and R. C. Stedman. "Positive Dependency and Virtuous Cycles: From Resource Dependence to Resilience in Urban Social-Ecological Systems." *Ecological Economics* 86 (2013): 292–299.

Tidball, K. G., and E. D. Weinstein. "Applying the Environment Shaping Methodology: Conceptual and Practical Challenges." *Journal of Intervention and Statebuilding* 5, no. 4 (2012): 369–394.

Tidball, K. G., E. D. Weinstein, and M. E. Krasny. "Synthesis and Conclusion: Applying Greening in Red Zones." In *Greening in the Red Zone*, ed. K. G. Tidball and M. E. Krasny, 451–486. New York: Springer, 2014.

Tocqueville, A. de. *Democracy in America*. Cambridge, MA: Harvard University Press, 1863.

Tuan, Y.-F. *Topophilia*. Englewood Cliffs, NJ: Prentice-Hall, 1974.

Tyack, N. "A Trash Biography." Los Angeles: Friends of the Los Angeles River, 2011. http://folar.org/wp-contentuploads/2011/11/trashsortreport.pdf, accessed July 3, 2014.

Tzoulas, K., K. Korpela, S. Venn, V. Yli-Pelkonen, A. Kazmierczak, J. Niemela, and P. James. "Promoting Ecosystem and Human Health in Urban Areas Using Green Infrastructure: A Literature Review." *Landscape and Urban Planning* 81 (2007): 167–178.

Ulrich, R. S. "View through a Window May Influence Recovery from Surgery." *Science* 224 (1984): 420–421.

United Nations Environment Programme. "Environmental Governance." 2009. http://www.unep.org/pdf/brochures/EnvironmentalGovernance.pdf, accessed February 23, 2014.

Urban Harvest. "Putting Down Roots across the City." 2014. http://urbanharvest .org/communitygardens, accessed August 2, 2014.

U.S. Department of Agriculture. "The People's Garden." http://www.usda.gov/wps /portal/usda/usdahome?navid=PEOPLES_GARDEN, accessed August 1, 2014.

U.S. Forest Service. "I-Tree." N.d. http://www.itreetools.org/, accessed July 3, 2014.

van der Steen, M., and J. Groenewegen. "Exploring Policy Entrepreneurship." In *Coherence between Institutions and Technologies in Infrastructures*, ed. M. Finger and R. Künneke (Rome: European Association for Evolutionary Political Economy, 2008), http://mir.epfl.ch/webdav/site/mir/users/181931/public/wp0801.pdf, accessed July 7, 2014.

Walker, B., S. Carpenter, J. Anderies, N. Abel, G. S. Cumming, M. Janssen, L. Lebel, et al. "Resilience Management in Social-Ecological Systems: A Working Hypothesis for a Participatory Approach." *Conservation Ecology* 6, no. 1 (2002), http://www.consecol.org/vol6/iss1/art14/, accessed July 7, 2014.

Walker, B. H., L. H. Gunderson, A. P. Kinzig, C. Folke, S. R. Carpenter, and L. Schultz. "A Handful of Heuristics and Some Propositions for Understanding Resilience in Social-Ecological Systems." *Ecology and Society* 11, no. 1 (2006), http://www.ecologyandsociety.org/vol11/iss1/art13/, accessed July 5, 2014.

Walker, B. H., and D. Salt. *Resilience Thinking: Sustaining Ecosystems and People in a Changing World*. Washington, DC: Island Press, 2006.

Wall, P. "Parkchester Block Beautification Group Branches out across the Bronx." September 4, 2013. http://www.dnainfo.com/new-york/20130904/parkchester/parkchester-block-beautification-group-branches-out-across-bronx, accessed February 26, 2014.

Wals, A. E. J., N. van der Hoeven, and H. Blanken. *The Acoustics of Social Learning: Designing Learning Processes That Contribute to a More Sustainable World*. Wagengingen, The Netherlands: Wagengingen Academic Publishers, 2009.

Wals, A. E. J., and M. E. van der Waal. "Sustainability-Oriented Social Learning in Multi-Cultural Urban Areas: The Case of the Rotterdam Environmental Centre." In *Greening in the Red Zone*, ed. K. G. Tidball and M. E. Krasny, 379–396. New York: Springer, 2014.

Warburton, J., and M. Gooch. "Stewardship Volunteering by Older Australians: The Generative Response." *Local Environment* 12, no. 1 (2007): 43–55.

Weber, E. P. *Bringing Society Back In: Grassroots Ecosystem Management, Accountability, and Sustainable Communities*. Cambridge, MA: MIT Press, 2003.

Weinstein, E. D., and K. G. Tidball. "Environment Shaping: An Alternative Approach to Development and Aid." *Journal of Intervention and Statebuilding* 1, no. 1 (2007): 67–85.

Wells, N. "At Home with Nature—Effects of Greenness on Children's Cognitive Functioning." *Environment and Behavior* 32 (2000): 775–795.

Wells, N. "The Role of Nature in Children's Resilience: Cognitive and Social Processes." In *Greening in the Red Zone*, ed. K. G. Tidball and M. E. Krasny, 95–110. New York: Springer, 2014.

Wells, N., and K. A. Rollings. "The Natural Environment in Residential Settings: Influences on Human Health and Function." In *The Oxford Handbook of Environmental and Conservation Psychology*, ed. S. Clayton, 509–523. Oxford: Oxford University Press, 2012.

Wenger, E., R. McDermott, and W. M. Snyder. *Cultivating Communities of Practice*. Cambridge, MA: Harvard Business School Press, 2002.

Wennersten, J. R. "The View from Watts Branch: Marvin Gaye Park Spurs an Environmental Transformation." May 2009. http://groundworkdc.org/wp-content/uploads/2010/08/Watts-Branch-Restoration_May09.pdf, accessed July 3, 2014.

Westphal, L. M. "Urban Greening and Social Benefits: A Study of Empowerment Outcomes." *Journal of Arboriculture* 29, no. 3 (2003): 137–147.

White, H. W. "Bergfelde: The Teacher and Conservationist Helga Garduhn Is Now 75 Years Old." Markische Online Zeitung, August 9, 2011.

Whyte, W. F., ed. *Participatory Action Research*. Thousand Oaks, CA: Sage, 1991.

Wilson, E. O. *Biophilia*. Cambridge, MA: Harvard University Press, 1984.

Wilson, M. A., and R. B. Howarth. "Discourse Based Valuation of Ecosystem Services: Establishing Fair Outcomes through Group Deliberation." *Ecological Economics* 41 (2002): 431–443.

Wimberley, E. T. *Nested Ecology: The Place of Humans in the Ecological Hierarchy*. Baltimore, MD: The Johns Hopkins University Press, 2009.

Winterbottom, D. "Developing a Safe, Nurturing and Therapeutic Environment for the Families of the Garbage Pickers in Guatemala and for Disabled Children in Bosnia and Herzegovina." In *Greening in the Red Zone*, ed. K. G. Tidball and M. E. Krasny, 397–410. New York: Springer, 2014.

Wolf, K. L. "Human Dimensions of Urban Forestry and Urban Greening." December 23, 2013. http://www.naturewithin.info/civic.html, accessed July 7, 2014.

Wolf, K. L., D. J. Blahna, W. Brinkley, and M. Romolini. "Environmental Stewardship Footprint Research: Linking Human Agency and Ecosystem Health in the Puget Sound Region." *Urban Ecosystems*, April 26, 2011. http://www.naturewithin.info/CivicEco/UrbanEcosystems.Stewardship.Apr2011online.pdf, accessed July 3, 2014.

Zelenski, J. M., and E. K. Nisbet. "Happiness and Feeling Connected: The Distinct Role of Nature Relatedness." *Environment and Behavior* 46, no. 1: (2012): 3–23.

# Index

In this index, the letter *b* following a page locator indicates information found in a box, *f* indicates a figure, and *t* indicates a table.

## Urban and Industrial Environments

Series editor: Robert Gottlieb, Henry R. Luce Professor of Urban and Environmental Policy, Occidental College

Maureen Smith, *The U.S. Paper Industry and Sustainable Production: An Argument for Restructuring*

Keith Pezzoli, *Human Settlements and Planning for Ecological Sustainability: The Case of Mexico City*

Sarah Hammond Creighton, *Greening the Ivory Tower: Improving the Environmental Track Record of Universities, Colleges, and Other Institutions*

Jan Mazurek, *Making Microchips: Policy, Globalization, and Economic Restructuring in the Semiconductor Industry*

William A. Shutkin, *The Land That Could Be: Environmentalism and Democracy in the Twenty-First Century*

Richard Hofrichter, ed., *Reclaiming the Environmental Debate: The Politics of Health in a Toxic Culture*

Robert Gottlieb, *Environmentalism Unbound: Exploring New Pathways for Change*

Kenneth Geiser, *Materials Matter: Toward a Sustainable Materials Policy*

Thomas D. Beamish, *Silent Spill: The Organization of an Industrial Crisis*

Matthew Gandy, *Concrete and Clay: Reworking Nature in New York City*

David Naguib Pellow, *Garbage Wars: The Struggle for Environmental Justice in Chicago*

Julian Agyeman, Robert D. Bullard, and Bob Evans, eds., *Just Sustainabilities: Development in an Unequal World*

Barbara L. Allen, *Uneasy Alchemy: Citizens and Experts in Louisiana's Chemical Corridor Disputes*

Dara O'Rourke, *Community-Driven Regulation: Balancing Development and the Environment in Vietnam*

Brian K. Obach, *Labor and the Environmental Movement: The Quest for Common Ground*

Peggy F. Barlett and Geoffrey W. Chase, eds., *Sustainability on Campus: Stories and Strategies for Change*

Steve Lerner, *Diamond: A Struggle for Environmental Justice in Louisiana's Chemical Corridor*

Jason Corburn, *Street Science: Community Knowledge and Environmental Health Justice*

Peggy F. Barlett, ed., *Urban Place: Reconnecting with the Natural World*

David Naguib Pellow and Robert J. Brulle, eds., *Power, Justice, and the Environment: A Critical Appraisal of the Environmental Justice Movement*

Roxanne Warren, *Rail and the City: Shrinking Our Carbon Footprint and Reimagining Urban Space*

Marianne E. Krasny and Keith G. Tidball, *Civic Ecology: Adaptation and Transformation from the Ground Up*